SpringerBriefs in Applied Sciences and Technology

SpringerBriefs present concise summaries of cutting-edge research and practical applications across a wide spectrum of fields. Featuring compact volumes of 50 to 125 pages, the series covers a range of content from professional to academic.

Typical publications can be:

- A timely report of state-of-the art methods
- An introduction to or a manual for the application of mathematical or computer techniques
- A bridge between new research results, as published in journal articles
- A snapshot of a hot or emerging topic
- An in-depth case study
- A presentation of core concepts that students must understand in order to make independent contributions.

SpringerBriefs are characterized by fast, global electronic dissemination, standard publishing contracts, standardized manuscript preparation and formatting guidelines, and expedited production schedules.

On the one hand, **SpringerBriefs in Applied Sciences and Technology** are devoted to the publication of fundamentals and applications within the different classical engineering disciplines as well as in interdisciplinary fields that recently emerged between these areas. On the other hand, as the boundary separating fundamental research and applied technology is more and more dissolving, this series is particularly open to trans-disciplinary topics between fundamental science and engineering.

Indexed by EI-Compendex, SCOPUS and Springerlink.

Luisa Lombardo

Smart Villages

Tools and Strategies for Sustainable Development in Inner Rural Areas

Springer

Luisa Lombardo
Department of Architecture
University of Palermo
Palermo, Italy

ISSN 2191-530X ISSN 2191-5318 (electronic)
SpringerBriefs in Applied Sciences and Technology
ISBN 978-3-032-16030-0 ISBN 978-3-032-16031-7 (eBook)
https://doi.org/10.1007/978-3-032-16031-7

This Springer imprint is published by the registered company Springer Nature Switzerland AG
The registered company address is: Gewerbestrasse 11, 6330 Cham, Switzerland

Preface

The theme of *Smart Villages* represents one of the most innovative, pressing, and fascinating challenges in contemporary territorial studies in Europe and around the world. Inner areas, often marginalized by processes of globalization and the centralization of rural services, are today confronted with the urgent need to redefine their development trajectories. Far from being merely "peripheral" spaces, these territories embody a complex heritage composed of traditional architecture, landscapes, built environments, customs, stories, history, and social networks that risk being overlooked if not adequately safeguarded and reimagined.

This work seeks to explore strategies for the rehabilitation and enhancement of inner rural areas through the smart village approach, understood not only as a technological framework but as a multidimensional model in which sustainability, governance, community participation, heritage conservation, and digital transition interact in a systemic and integrated way. Its aim is to show how the principles of smart cities can be reinterpreted and adapted to rural contexts, providing both practical tools and theoretical frameworks for sustainable development.

The book is organized around three main research directions: first, the analysis of the theoretical background of smart villages and their relationship to the broader discourse on sustainability and innovation; second, the examination of selected case studies, drawn from diverse contexts in Europe and beyond, which illustrate a range of strategies and approaches to rural regeneration; and third, the development of operational tools conceived as a practical guide to interventions that respect both the fragility and resilience of inner areas, balancing tradition and innovation.

By addressing these themes, the book engages with fundamental questions at the heart of the rural regeneration debate: how can innovation be reconciled with tradition? How can heritage be safeguarded while enabling new forms of development? And ultimately, how can inner areas become laboratories of experimentation for sustainable futures?

In exploring these questions, the chapters provide insights not only for specific case studies but also for broader applications, offering a reference framework for policymakers, practitioners, and scholars interested in promoting smart, sustainable, and resilient rural communities. The ambition of this work is to contribute to the construction of a shared vision of smart villages as spaces of opportunity, where the rehabilitation of heritage and the valorisation of territorial resources align with the needs of contemporary society.

Palermo, Italy Luisa Lombardo

Acknowledgements I would like to express my deepest gratitude, first and foremost, to my supervisor, Prof. Tiziana Campisi, an authoritative and unwavering guide throughout these three years of research. Her expertise, dedication, and academic passion have been a constant source of inspiration, particularly during the most challenging moments of my work. Her support has been a fundamental reference point in my journey, profoundly shaping both my research and professional growth.

I am also deeply grateful to Prof. Simona Colajanni, who has accompanied my professional development over the past ten years with generosity, invaluable guidance, and steadfast encouragement. Her support has been essential to my formation, and without her belief in me, I might not have come this far.

My sincere thanks go to Prof. Manfredi Saeli for his dedication and for providing me with the opportunity to experiment with innovative materials in the Madonie area, monitoring their performance and expanding my research horizons. His guidance has been crucial in exploring new technological solutions for the Madonie Park.

I owe heartfelt thanks to my family for their unconditional support and tireless assistance, including during the many field trips to the Madonie, which allowed me to pursue my research with focus and determination.

I would like to acknowledge Prof. Arturo Zaragozá Catalán and Prof. Rafael Marín Sánchez for introducing me to Valencia, its architecture, and its cultural heritage, and for their invaluable contribution to the development of the Guidelines for the Rehabilitation of the Municipality of Polizzi Generosa.

I am also grateful to Architect Delia Geraci for her support in drafting the Atlas of the Madonie Area, a critical tool for identifying the most appropriate technological solutions within this well-preserved context.

Special thanks are due to all the municipalities of the Madonie UNESCO Geopark and the Municipality of Polizzi Generosa, including Mayor Gandolfo Librizzi, Head of the Technical Office Engineer Valentina India, and Mr. Enrico Cerami, whose local collaboration and support were indispensable for the study of the municipality and the implementation of projects, events, and technological facilities.

I extend my gratitude to the PG 5 Cuori Foundation for embracing and funding the proposed solutions, including the totems and lighting of the "tocchi" of Polizzi. My thanks also go to the Borgese Foundation and its President, Gabriella Clara Aiosa, for their trust and collaboration in presenting the Borgese project to the PAC Call from the Ministry of Culture, and for their support in hosting the Summer School and the Smart Village proposal in the heart of the Madonies.

I am grateful to the Municipalities of Petralia Sottana and Isnello, their mayors, councillors, and local staff, for their openness, collaboration, and support in the development of the Summer School and Smart Village initiatives. I also wish to thank the Madonie community for their curiosity, initiative, and willingness to engage with my research.

I sincerely appreciate the contributions of all the speakers at the seminar on Energy Communities, including Chiara Onorati, Dr. Lucilla Parisi, Dr. Vincenzo Raffa, Councillor Carmelo Bellomo, Dr. Angelo Conti, Engineer Andrea D'Amore, and Engineer Matteo Piazza. Their insights and discussions greatly enriched the analysis of projects and opportunities in the Madonie territory.

Finally, special thanks go to Prof. Giuseppe Trombino, Dr. Mario La Rocca, Prof. Montserrat Bosch González, Dr. Alessandro Ficile, Dr. Calogero Puleo, Dr. Luciana Cusimano, and Mario Puglisi, whose contributions allowed for a deeper exploration of energy community development within the Madonie Park.

I am also grateful to my Ph.D. colleagues for their stimulating discussions and continuous support, which have been fundamental to both my professional and personal development. My thanks extend to Samuele Morvillo, GiN APS, the students of AISA, and to Adriana Calà and Martina Famoso for their encouragement and attentive listening. Finally, I wish to thank the entire Scientific Commitee of Curriculum in Sustainable Architecture and Design: Human Centered Approach (from the 37th cycle) for supporting the project and contributing to a robust and methodologically sound research process.

This book and its findings are the result of the dedication, collaboration, and support of all these individuals and institutions, whose contributions have made the research on the Madonie both tangible and meaningful, offering insights and inspiration for rural areas and the future of rural development.

Competing Interests The author has no competing interests to declare that are relevant to the content of this manuscript.

Contents

Chapter 1
The Development of Inner Areas Through the Creation of Smart Villages, a Shared Project

Abstract This chapter provides a comprehensive analysis of Smart Villages at global, European, and national levels, highlighting strategies to support the sustainable development of rural areas. It focuses on digital transition, technological innovation, and new governance approaches as key drivers for resilient and competitive local communities. Selected international and Italian case studies illustrate how sustainable solutions, digital tools, and community participation contribute to the revitalization of inner areas. The chapter also examines European and national policies, initiatives, and funding instruments that can support small municipalities, emphasizing alignment with broader strategies such as the European Green Deal and Agenda 2030. The combination of case study insights and policy frameworks establishes a solid foundation for understanding and applying Smart Village strategies.

1.1 SMART Criteria: Defining and Achieving Goals with Precision and Success

In the modern era, technological evolution has led to a profound change in the conception of cities, giving rise to *Smart Cities*. These have developed greater intelligence, transforming in response to the need to adapt to the present. Initially, they were conceived as urban centres where information and communication technologies were integrated to improve the quality of life for citizens. This approach led to the optimization of public services, efficient resource management, and increased environmental sustainability. The term "*Smart City*," which has gained popularity in recent years, refers to cities that use technology and innovation to improve efficiency, service quality, and sustainability, while also focusing population in areas where services are greater and more efficient. However, rapid urbanization and the concentration of resources in large cities have also highlighted significant challenges, such as pollution, traffic congestion, and social stress. These challenges have prompted an expansion of perspective, leading to the conceptual evolution from *Smart City* to *Smart Land* [1].

L. Lombardo, *Smart Villages*,
SpringerBriefs in Applied Sciences and Technology,
https://doi.org/10.1007/978-3-032-16031-7_1

The transition to *Smart Land* reflects a broader vision that considers not only cities as epicentres of innovation but also rural areas as an integral part of a socio-technological ecosystem [2]. This shift in perspective is driven by the goal of creating a balance between urban and rural development, embracing a range of key concepts, including:

- rural regeneration: it promotes the rehabilitation and enhancement of rural areas, which have often been neglected in favour of cities. Through investments in digital infrastructure, smart agriculture, and sustainable tourism, rural communities can regenerate, becoming centres of innovation and development [3];
- smart rural villages: they represent a new approach to living, where technology is used to improve the quality of life in rural contexts. Access to advanced digital services, efficient energy management systems, and sustainable mobility solutions can transform these villages into vibrant and resilient communities [4];
- widespread connectivity: a fundamental element that reduces the digital divide between cities and rural areas. Access to high-speed internet and mobile coverage is essential to ensure that rural communities can fully benefit from technological opportunities [5];
- integrated sustainability: promoting eco-friendly agricultural and industrial rehabilitation practices and encouraging the conservation of the natural environment and traditional architecture. This creates a balance between economic growth and environmental preservation [6];
- circular economy: waste produced is minimized, reused, and resources are used more efficiently, fostering the creation of self-sufficient communities [7] not only in terms of disposal but also their reuse in daily life, such as in the construction field.

The evolution from the concept of *Smart City* to that of *Smart Land* represents a step forward in building a sustainable and inclusive future. The promotion of smart rural villages and the enhancement of rural areas not only balance urban development but also embrace the very essence of the connection between humans and the environment, creating a balance between the advantages and disadvantages that characterize them. This transition represents a new frontier of living, where technology, sustainability, and community converge to create a more balanced and harmonious world [8]. The metamorphosis from *Smart City* to *Smart Land*, first proposed by Prof. Roberto Masiero in 2014 [1] for the Italian context, represents an innovative and significant approach in the field of territorial development and opens new opportunities for the small marginal areas of our Country. The concept reflects a shift in perspective, moving the focus not only from cities but also towards rural areas and less densely populated territories, such as the outskirts or the inner villages of the Alpine and Apennine regions. This transition, which is widespread throughout Europe, aims to promote harmonious and balanced development, embracing the entire geographical and social landscape, without excluding any part of it [9]. It is, in fact, equally crucial to pay attention to small urban centres on the outskirts, along with rural areas, which constitute a significant part of the Italian and European territory. Thanks to technology and increased environmental awareness, wide opportunities have opened for the smart development of these territories. It has emerged that digital innovation can bring considerable benefits to rural centres, enhancing services and preserving distinctive features, combining tradition and innovation. It is also important to note that these European inner areas host around 93 million citizens, making it essential to promote smart development.

Masiero argues that focusing solely on urban development could lead to an imbalance between urban and rural areas, risking the marginalization of the latter. The proposal of *Smart Land*, however, is based on the idea that technological innovation should spread uniformly across the entire territory, including both cities and rural areas [10]. This approach aims to leverage the potential of rural communities, preserving their traditions and enhancing local resources through the adoption of technological and innovative solutions. Cities thus become a fundamental support for small urban centres, bridging the existing *gap*. They thus help in the enhancement of local resources and heritage in general, preventing overcrowding in certain centres, now defined as critical, compared to others that are less populated and subject to depopulation. However, this involves several aspects, including:

Provide access to information and communication technologies (ICT) such as high-speed connectivity and *IoT* (Internet of Things) solutions [11];

- apply technological solutions to promote environmental sustainability, such as the use of renewable energy and eco-sustainable agricultural practices, enabling rural communities to develop in an eco-compatible manner;
- enhance the traditions and cultural heritage of rural areas, preserving their unique identity while embracing innovation;
- implement sustainable mobility solutions in rural areas, such as eco-friendly public transport and infrastructure for soft mobility, contributing to accessibility and connectivity;
- actively involve local communities in the development of strategies and projects, promoting collaboration and the sharing of ideas;
- foster the development of new local businesses and job opportunities, utilizing existing resources and creating new economic prospects.

The perspective of the *Smart Land* represents a great opportunity to reduce imbalances between urban and rural areas, creating a more balanced social and economic fabric. Promoting innovation and the adoption of technologies in all territorial regions could lead to sustainable and inclusive progress for the entire national community, and this model could be replicated and expanded both in Europe and overseas. This concept of *Smart Land* takes shape, later in Europe, in the form of *Smart Villages* [9], which combine information and communication technologies (ICT) with the picturesque and remote rural areas, villages, smaller towns, and countryside. To better understand them, we can simplify the concept by comparing it to the idea of a *Smart City*, where technology and digital innovation enhance sustainability, services, information, and the use of innovative strategies [12]. The goal of *Smart Villages* is to improve various aspects related to daily life by expanding the possibilities of rural areas, including energy efficiency, connectivity, access to services, resilience, security, environmental protection, citizen participation, and the adoption of new sustainable business, work, and mobility models, supported by active and essential local participation [13].

The industrial revolution of the twentieth century, as well as the COVID-19 pandemic, have re-evaluated S*mart Villages* since 2020. This invites reflection on their promotion, fostering a high quality of life and adopting a circular economy instead of indiscriminate global resource consumption. The goal of this paradigm shift is to promote the endogenous resources of the territory, which are more abundant in these areas, and to offer economic value that is currently largely absent or still underappreciated [14]. However, it is essential to properly redevelop and revitalize rural villages that have fallen behind in terms of progress. This includes strengthening infrastructure and telecommunications, network coverage, expanding *Wi-Fi* zones, producing sustainable energy, ensuring sustainable mobility, and revitalizing existing structures—lands that have been left behind in time and victims of the encroachment of large cities. These changes inevitably lead to the strengthening of the tourism sector, considering new strategies like smart working, which is key to increasing both demand and supply. This could, for example, transform vacation homes into true residences for remote workers or, even better, digital nomads. Similarly, the concept of living and urbanity changes, encouraging investment to shift toward peripheral areas to create a network of *smart villages* that attract labour, professionals, families, and businesses.

Cutting-edge communities that foster collaboration and connections, involving various *stakeholders*, both public and private, and promoting projects that would otherwise be unachievable, without necessarily having a specific administrative connotation [15]. Unlike cities, these can benefit from collaboration between citizens and small towns to gain the same urban opportunities; it is well known that within small territorial contexts, there is a much closer and more familiar neighbourhood relationship, essentially a large condominium where the principle of mutual collaboration prevails. The projects to be undertaken must, therefore, vary based on local needs, understand territorial vocations, and develop them, allowing for targeted interventions to address issues such as abandonment and promoting repopulation as an alternative to urban chaos. Investing in the development of connected rural networks and *Smart Villages* is of fundamental importance not only for the Italian context but also for the European one. These areas make up more than half of the European territory, are inhabited by millions of citizens who deserve services and a quality of life comparable to those offered in large cities, and in Italy, they cover more than 60% of the country, hosting a significant portion of the population [16].

To achieve sustainability goals by 2030 and 2050, it is crucial to focus on programs that make a significant contribution to the environment and sustainable development, providing greater flexibility and the opportunity to experiment with new innovative solutions without compromising the landscape and environment. Efforts can be made across various aspects, including the economy, environment, *governance*, lifestyles, mobility, and technology, to promote sustainable development. These are valuable resources that, if properly utilized, can contribute to the transition towards a more sustainable future [17].

Investing in digital connectivity, renewable energy, smart agriculture, and other innovative solutions can promote economic growth, job creation, environmental protection, and improved quality of life in rural communities. Adopting a holistic approach that involves a wide range of *stakeholders*, including local governments, businesses, academic institutions, and the community itself, is essential for developing targeted strategies and projects [18]. The goal, however, is to create *Smart Villages*, networks of villages that keep pace with the times, capable of offering efficient, sustainable, and inclusive services, improving people's lives while preserving the cultural and environmental identity of rural communities [19], just like the comfortable city that was aimed for with the *Smart City* concept. Dynamic villages where change and technology, strategies and innovations go hand in hand. Innovative solutions to enhance resilience, starting from local strengths and opportunities, with an approach that involves the community in developing strategies to improve economic, social, and environmental conditions. Villages that benefit from collaboration with other communities, both urban but especially rural, and that can be funded both publicly, drawing from European funds, and privately, due to their potential and the principle of mutual collaboration, distinct from unknown citizens. They represent a significant opportunity for improving quality of life, developing a more sustainable and resilient economy [20]. In *Smart Villages*, it is therefore possible to implement a wide range of specific projects, depending on local needs and community characteristics [21], including examples of such projects.

- digital infrastructures: Implementation of high-speed networks and Internet connectivity in rural areas to enable access to digital technologies, promote e-commerce, improve remote education, and facilitate telemedicine, especially due to the frequent lack of healthcare facilities;
- sustainable energy: Promotion of renewable and smart energy solutions, such as the installation of solar panels, the use of microgrids, and energy storage systems, to reduce environmental impact and improve energy efficiency in rural communities, ensuring *smart* and innovative buildings;
- smart agriculture: Adoption of advanced agricultural technologies, such as precision farming, the use of sensors for crop monitoring, automation of agricultural processes, and the implementation of smart irrigation systems to enhance productivity and agricultural sustainability, as well as increased production;
- sustainable mobility: Promotion of sustainable transportation solutions, such as carpooling, bike-sharing, electric vehicles, and the implementation of electric vehicle charging infrastructure, to reduce pollution and improve mobility in rural areas;
- smart services: Development of advanced digital public services, such as e-government, smart waste management, efficient street lighting, implementation of smart security systems, and the use of home automation solutions to enhance the quality of life in the community;

- sustainable tourism: Enhancement of the cultural and natural heritage of rural areas using digital technologies, creation of innovative tourist experiences, implementation of sustainable tourism management systems, and promotion of low environmental impact tourism activities;
- local economy and entrepreneurship: Support for innovation and local entrepreneurship through the implementation of coworking spaces, organization of training programs and support for *startups*, promotion of local production chains, and the creation of collaboration networks between businesses;
- heritage rehabilitation and enhancement: Promotion of traditional architecture considering innovation and technological compatibility.

These are just some of the themes and ideas that can be implemented to promote *Smart Villages*, and this work aims to analyse them one by one. However, the specificity and diversity of each depend on the needs and resources of the individual communities, as well as the willingness to collaborate among the various *stakeholders* involved. It is important to emphasize that not all the strategies listed above can be adopted in all contexts, but they can be versatile and promoted in different ways [22] based on their territorial compatibility. To continue the previous studies, in these years of research, policies supporting "*smart villages*" have been examined, considering international, European, and Italian strategies, and concrete case studies have been provided for a comprehensive view of the work done so far. These studies can find innovative and effective solutions to pursue with solid foundations and a well-defined path, considering various aspects. In the following paragraphs, the theme of *Smart Villages* will be examined as an introduction to fully understand the summary sheets, explanatory examples, and case studies that will support the doctoral research.

"*Future cities are neither the invisible ones of Italo Calvino nor the visionary ones of Fritz Lang. They are the cities of today, seen and lived in through a more efficient perspective—technological and creative at the same time, and above all, participatory.*" Mattei, M.G. (2013). *Carlo Ratti. Smart city, smart citizenz.* Milano: Egea Edizioni.

1.2 Non-European Smart Villages

In the current context, where just over half of the global population has access to the internet while the other half remains disconnected, the global COVID-19 pandemic has highlighted the vital importance of digital networks and services for economic resilience and the continuity of essential public services such as education and healthcare. In this scenario, collaboration and multi-*stakeholder* partnerships are critical for reaching underserved and unconnected communities and for pursuing the goal of universal and meaningful connectivity. It is particularly in rural areas that this need is most apparent. With this perspective, the International Telecommunication Union (ITU) and its partners have promoted the creation of *smart villages* to foster sustainable development in remote and disadvantaged areas of the world. The first *Smart Villages*, in fact, emerged outside Europe, particularly in developing countries such as Africa, India, Indonesia, and Nepal. These regions, where growth opportunities are limited and much work remains to be done, provide a starting point from which to build from the ground up. The term *Smart Village* emerged around 13 years ago with the idea of promoting the development of areas that, for us, might be considered small rural communities, but which for economically disadvantaged areas are collections of *Smart Villages* united by the lack of essential goods and services needed to ensure a dignified life for their inhabitants. In response to the desire to assist those who have less than we do, numerous non-profit organizations, associations, and even large multinational corporations have taken action to help these more vulnerable communities.

While the term "SMART" is often associated with the digital world, observing the various projects undertaken worldwide, it becomes clear that it is also linked to the use and exploitation of natural resources to create new businesses and promote local products, facilitating entry into the large international market and fostering social and economic development practices. This primarily involves using digital tools to promote culture and innovation, provide education for women, ensure sustainable resource management, harness renewable energy sources, and, finally, promote a local identity that is often unknown. Therefore, the term "*smart*" is associated with the wise use of digital intelligence to activate virtuous processes of local development, which are based not only on significant financial investments but also on a huge and indispensable investment of human capital. What truly makes these villages *smart* is the collaboration among the resident population, which provides the added value necessary to achieve sustainable and concrete parameters of liveability and dignity. The unity among citizens represents the true success, further driven by the desire to surpass their own capabilities and potential, and most importantly, to achieve a sense of empowerment with the goal of attaining better and more satisfying long-term results, to be on par with developing countries that are significantly more advanced. Exploring a series of *Smart Villages* identified worldwide, excluding those in Europe and Italy due to significantly different policies and *governance*, successful examples emerge particularly in the United States and Asia, but also in other parts of the globe. The starting point is not population shortage, but rather the lack of services and sometimes widespread poverty. A *Smart Village* can also be highly populated; it is the lack of opportunities and the willingness for technological implementation that promotes shared territorial development.

In the United States, for example, *Smart Villages* have spread across various regions with the goal of revitalizing rural communities and promoting sustainable development. For instance, *Chattanooga* in Tennessee is famous as one of the first smart locations in the United States, thanks to the implementation of high-speed broadband networks and internet access in the surrounding rural areas, which fostered the growth of tech businesses and the creation of jobs in the information and communications sector. Fresno, California, adopted a smart farming approach by leveraging the Internet of Things (*IoT*) to monitor agricultural data and optimize irrigation, thereby reducing water consumption and increasing crop yields.

Jefferson, Iowa, became an example of a *Smart Village* through the use of renewable energy sources such as wind and solar power, contributing to the reduction of carbon emissions and promoting energy sustainability (*Rural Broadband Access Loan and Loan Guarantee Program—RBALGP; Rural Development Broadband Initiative (RBDI),—Rural Utilities Service (RUS)*: an office for loans, grants, and other financial assistance to help rural communities provide public services, including electricity, water, natural gas, and broadband). In Asia, for example, *Kerala* in India has implemented a *Smart Village* model aimed at improving access to healthcare and education using digital technologies, thus enhancing the accessibility and quality of services through programs such as the *National Smart Village Construction Plan (2014–2020)* and the *National Smart City Construction Plan* (2014–2020). Another significant initiative is represented by *Barefoot College*, an Indian project extended to several African countries, such as Kenya, Burkina Faso, Senegal, and others. This organization has adopted a *Smart Village* model focused on the training and education of women in rural communities, known as "*solar mamas*", to build and maintain solar panels, thereby providing clean and sustainable energy to their communities. In Japan, *Smart Villages* are often linked to the concept of "Society 5.0," which aims to combine advanced technologies with human well-being. Japanese rural communities are adopting innovative solutions to address the aging population, such as the use of robots for elderly care and the implementation of smart systems for monitoring health conditions, thanks to the 2019 *Smart Rural Community Development Program*. In Africa, *Smart Villages* represent an important initiative to tackle the challenges faced by rural communities on the continent and promote sustainable development at the local level. Among the examples of *Smart Villages* in Africa is the *Re:learn Hub of CcHub* (Co-Creation Hub) based in Lagos, Nigeria. This project focuses on the use of information and communication technologies (ICT) to improve education in rural communities, providing spaces and resources for digital learning and teacher training, with the goal of reducing the digital divide that often characterizes these areas. In South Africa, there is the Digital Villages initiative, which aims to bring digital technologies to the country's rural communities, thus improving access to essential services such as education, healthcare, and legal advice. This initiative works towards the construction of community centres equipped with internet access, computers, and other digital resources (African Union Smart Villages Program 2018 and International Monetary Fund Smart Villages Program 2020). Additionally, it is important to mention the *Smart Africa* initiative, which represents an ambitious pan-African effort aimed at promoting the development of information and communication technologies (ICT) across the continent. This initiative focuses on improving access to ICT, encouraging innovation, and promoting socio-economic development, including the implementation of similar projects in various African countries.

The initiatives mentioned above represent just a few examples of how digital and innovative technologies are helping to transform rural communities in non-European territories and address the challenges of rurality with sustainable and innovative solutions. The success of *Smart Villages* outside Europe has been achieved partially and is still under development, thanks to several factors, including the strong synergy between public and private actors (the main economic supporters of such initiatives), investments in digital infrastructure, the active involvement of local communities, and great collaboration with multinational companies, universities, and research institutions. The strong participation of communities ensures long-term sustainability and the well-being of those involved, demonstrating that non-European *Smart Villages* can inspire Europe to pursue innovative strategies for the future of rural areas, promoting sustainability, innovation, and economic prosperity beyond "*top-down*" funding. In the following pages of this section, an in-depth analysis of non-European *Smart Villages* will be conducted. The focus will be on various regions of the world, particularly examples from the Americas, Asia, and Africa, as outlined above, but not limited to them. This will highlight how *Smart Villages* represent an innovative approach to addressing the complex socio-economic and environmental challenges that rural communities face in different parts of the world. Through the strategic adoption of digital technologies and the implementation of sustainable approaches, these projects aim to improve the quality of life, stimulate economic development, and carefully preserve the cultural identity of the rural areas involved. Examples of *Smart Villages* will be explored where various initiatives have been launched to revitalize rural communities and capitalize on the opportunities offered using digital and sustainable technologies (Fig. 1.1). Through a thorough analysis, the goal is to provide a comprehensive overview of the many ongoing initiatives worldwide and to draw *insights* and inspiration for successfully addressing the global challenges faced by rural communities, pursuing a sustainable, inclusive, and respectful future for cultural and environmental identities, as should be pursued equally in Europe and around the world.

1.3 European Smart Villages

European *Smart Villages*, like those outside Europe, represent an innovative approach aimed at revitalizing rural areas through the implementation of digital technologies and sustainable solutions, as already mentioned for the extra-European context. By leveraging digitalisation and sustainable practices, European *Smart Villages* promote social, economic, and environmental development, paving the way for vibrant and resilient rural communities in Europe. They embrace digital technologies, sustainable practices, and community engagement to improve quality of life and foster economic growth. Key elements include the priority of digital connectivity, sustainable energy solutions, innovative services, entrepreneurship, job creation, as well as social and community involvement in depopulated areas, the improvement of quality of life through enhanced services, promotion of environmental sustainability, and empowerment of rural communities. They primarily contribute to combating depopulation, attracting and retaining residents through job opportunities and an environment conducive to well-being. However, inner European countries face several challenges, including limited resources, regulatory barriers, resistance to change, and the need to ensure inclusion and digital literacy in rural areas. Additionally, there are significantly fewer grants from private sources, compared to public ones.

Fig. 1.1 **A** Chattanooga, Tennessee © Tripadvisor, 2023; **B** Kerala, India © Saga Magazine, 2023; **C** Barefoot College, Africa © The New York Time, 2011; **D** Samsung Digital Village, Sud Africa © ITWeb Africa, 2015

At the political level, the European Union [23] provides strong support through several instruments such as the Cohesion Policy, the European Broadband Strategy, the Digital Agenda for Europe, the Common Agricultural Policy and the Horizon Europe Program. Furthermore, the European *Smart Villages* Initiative has been established to coordinate and promote the implementation of policies supporting *smart villages* across Europe, facilitating the exchange of knowledge and *best practices* between regions. European policies supporting *Smart Villages* are a key pillar for the revitalization of rural areas through innovation and digitalisation [20]. Through funding, guidelines, and strategic initiatives, the EU promotes digital connectivity, sustainability, and social inclusion in *Smart Villages*. Unlike extra-European examples, the European Union is significantly focused on promoting *smart villages*, prioritizing available public funds over privatized initiatives [24]. These important policies provide opportunities for economic development, job creation, environmental preservation, and improved quality of life in rural European areas. The EU's ongoing commitment to supporting *Smart Villages* plays a substantial role in creating resilient, innovative, and future-oriented rural communities within the European landscape.

The European Network for Rural Development (ENRD), for example, represents a collaboration platform that brings together *stakeholders* from across Europe to promote innovation and sustainability in rural areas. Founded with the aim of facilitating the sharing of knowledge, experiences, and *best practices*, it plays a crucial role in supporting the planning and implementation of EU rural development policies. *Smart Villages* in Europe represent an increasingly widespread and diverse reality, present in various countries across the continent. Some of the best examples in Europe include [25]. In Italy, *Damanhur* in Vidracco, Piedmont, is an intentional community located in the northwest of Italy. It is known for creating an underground system of temples and artistic galleries, as well as for adopting ecological and sustainable practices. Damanhur has also developed a cooperative organization for resource management and democratic participation of its residents. *Marmilla Smart Community*, in the Marmilla geographical subregion in Sardinia, is a project focused on using digital technologies to improve local services, promote sustainable tourism, and support the circular economy, thus contributing to the well-being of the local community. *Borgo Soandri*, in the Marche region, is an ancient rural village that has been redeveloped as a *Smart Village*, with particular attention to environmental sustainability and efficient resource use through the implementation of a diffused hotel aimed at encouraging eco-sustainable tourism. Elsewhere in Europe, in Portugal: *Aldeias Do Xisto*, located in central Portugal, is a project that transformed 27 abandoned villages into sustainable tourist destinations. Through the revitalization of historic buildings, the use of renewable energy, and offering authentic cultural experiences, the project has contributed to revitalizing rural communities and promoting sustainable tourism. *Tamera*, a sustainable village located in Alentejo, was designed as a model of a sustainable community based on energy self-sufficiency, organic farming, and water resource management. Tamera is also committed to research and education on topics such as peace, ecological healing, and cooperation among people. *Aldeia Solar* is known for its use of renewable energy sources and sustainable practices to reduce environmental impact and promote sustainability.

In Spain: *Lavafante*, in the Asturias region, has adopted innovative technologies for the sustainable management of water resources. Using sensors and smart irrigation systems, Lavafante has reduced water consumption in agriculture and implemented sustainable farming practices to preserve the natural environment. *Mata De Alcǽntara* focuses on energy self-sufficiency, smart mobility, and the digitization of public services, aiming to improve the quality of life for residents in this rural area. *Urdaibai*, in the Basque Country, is a biosphere reserve that has implemented sustainable solutions for resource management and the promotion of responsible tourism.

In Denmark: *Samsø:* an island known as one of the first *Smart Villages* in the world to be powered by renewable energy, achieving energy self-sufficiency using wind and solar power, as well as promoting energy efficiency practices. It has now become a popular tourist destination and has created jobs in the renewable energy sector.

In Austria: *Schlierbach*: a village that has successfully implemented smart solutions for energy and water efficiency. Using advanced technologies such as energy monitoring and building management systems, it has reduced energy consumption and promoted the sustainable use of water resources.

In the Netherlands: *Schoonschip*, is a sustainable floating village where homes are equipped with innovative technologies for energy efficiency and the sustainable use of water resources.

In Germany: *Schwerin*, in the Mecklenburg-Vorpommern region, has adopted smart technologies for building energy efficiency, waste management, and sustainable mobility. *Freiburg Vauban* is a model for sustainable urban development that highlights ecological technologies, energy efficiency, and sustainable mobility, with a focus on environmental impact.

In the United Kingdom: *Winterton-On-Sea*, near Norfolk, is a community that has adopted renewable energy, such as solar and wind power, to power its buildings and reduce environmental impact.

In France: *La Ferme Du Bec Hellouin*, in Normandy, is an organic farm that uses sustainable and intentional farming techniques to grow food ecologically.

In Sweden: *Hjulsjö*, Närke, has experimented with the use of smart technologies to optimize energy use and reduce waste. *Österbymo Smart Village* is a digitally connected community offering high-speed internet access, smart grid solutions, and innovative services, aimed at improving the lives of its residents.

In Finland: *Oulu Innovation Village*, in Oulu, is a technological and entrepreneurial innovation area that promotes collaboration and the development of smart solutions; *Lyseparken*, is a neighbourhood in Stavanger that integrates smart solutions for public lighting, waste management, and sustainable mobility; *Brütten*, in Zurich, is a Swiss municipality that has adopted ecological technologies for energy supply, such as solar power and cogeneration.

In Switzerland: *Tänikon Smart Village* uses smart farming techniques, renewable energy systems, and electric mobility solutions to create a sustainable rural community at the forefront of innovation.

In Estonia: *Lõunakeskuse Smart Village*, in Tartu, focuses on using digital and sustainable solutions to improve daily life in the community and make the village a modern and well-organized place.

In Norway: *Gjesdal Smart Village* uses digital technologies to improve waste management, mobility, and actively engage citizens in village life.

In North Macedonia: *Belasica Smart Village* has the main objective of using digital technologies to promote sustainable tourism and economic development in the region [26].

These are just a few of the *Smart Villages* in Europe that have demonstrated success in applying digital technologies and sustainable solutions for rural community development, thanks to funding from Europe [27]. It is important to note that these examples represent only a small selection of the numerous, ever-evolving initiatives that are radically transforming rural areas in Europe. Each *Smart Village* is unique and responds to the specific needs and challenges of its community and territory, contributing to the creation of a more sustainable and prosperous future for the rural areas of the continent. Within the broader context of *Smart Villages*, there are numerous European projects that have addressed this topic. However, the most important are *Smart Rural 21*, *Smart Rural 27*, and the pilot project "*Eco-Social Smart Villages*". In particular, the *Smart Rural 21* and *Smart Rural 27* projects represent two initiatives funded by the European Commission with the common goal of promoting the development of *smart villages* in Europe through selected pilot cases. Below is a formal description of both projects: *Smart Rural 21*, a project lasting two and a half years (2018–2020), it involved 21 pilot villages in 12 EU member states. Its mission was to develop and implement strategies aimed at transitioning these villages into *smart communities*. The specific objectives included social innovation, collaborative *governance*, local economic development, and environmental sustainability. *Smart Rural 27*, a three-year project extended from 2020 to 2023, it involved 27 EU member states. The main objective was to assist member states in defining frameworks for creating infrastructure supports for *smart villages*. The key focus areas included policies and strategies for rural development, capacities and skills, funding, and investments. Both projects highlighted that being a *smart village* involves the use of digital technologies and innovation to address local challenges, improving the quality of life for residents. These villages rely on innovative and sustainable approaches, incorporating technological solutions to optimize services, infrastructure, and resources, with the aim of creating more resilient and future-oriented communities.

At the same time, it is worth mentioning the pilot project *Smart eco-social villages*, summarized in the executive report drafted by the *Directorate-General for Agriculture and Rural Development of the European Commission*, which identifies *best practices* to support the future development of strategies. The project is structured around four main themes, analysing opportunities and challenges in the rural areas of the European Union, defining *Smart Villages*, illustrating examples of *best practices*, and exploring the development of a strategy for a smart village. The document emphasizes that *Smart Villages* are rural communities that use innovative solutions to improve their resilience, leveraging digital technologies and promoting active participation from the local community, as well as the importance of effective *governance*, active community involvement, simplicity in strategies, and cooperation with other rural and urban communities and actors. Furthermore, it stresses the need to simplify and streamline procedures for accessing public funding, support small-scale investments, and promote cooperation between communities. The project underscores the importance of providing support at the European, national, and regional levels for the development of *Smart Villages*, with particular attention to less favoured rural areas. The three key lessons provided are the wide variety of initiatives in rural EU areas, the importance of citizen participation, and the need for a flexible and targeted support framework across different levels of government. It thus offers a detailed overview, providing the first important guidelines for supporting and developing such initiatives in rural areas of the European Union. By analysing the challenges and opportunities faced by rural areas in the EU, the pilot project highlights the need for simplified public funding policies and support at different levels of government to foster the development of *Smart Villages* in the European Union [25].

A roadmap that can be adapted to the specific needs of the study area. The implementation could constitute a significant step towards a coherent and integrated policy capable of addressing territorial challenges more effectively. An appendix will focus on additional case studies of *Smart Villages* in different European countries (Fig. 1.2) [28]. Each case study, as was done for the analysis of non-European contexts, will provide a broad overview of the characteristics and objectives of each project, highlighting the results achieved and the innovative solutions adopted. These examples represent just a selection of the many initiatives contributing to the transformation of rural areas in Europe, but they will provide a comprehensive framework of the potential and benefits of implementing such practices in the European context. A thorough understanding of these case studies is crucial to fully grasp their positive impact across Europe. Each project tells a unique story of resilience, innovation, and community involvement, demonstrating how digitalisation and sustainability can contribute to a more prosperous future for these European rural areas. Through these examples, the hope is to inspire other communities to undertake similar initiatives and provide valuable *insights* for technological and social progress. The analysis presents new opportunities and challenges for European rural areas, where the implementation of smart and innovative solutions appears to be a shared hope.

1.4 Italy Towards the Development of Smart Villages

In an increasingly homogeneous and globalized world, the forgotten towns, especially those in the Italian hinterland, often labelled as "villages," represent a precious and authentic part of our cultural and social heritage. However, the time has come to abandon the reductive label of "village" and embrace the more inclusive and meaningful term "town." Italian towns are much more than simple urban agglomerations; they are the beating heart of our regional and national identity. These places, which can be densely populated or suspended in an apparent calm, under the sun or shrouded in fog, reveal a variety of stories, traditions, and lifestyles that reflect the richness of our nation. They are communities of people who know each other, help one another, and build the social fabric that forms the foundation of our society.

While the "village" concerns the urban dimension and defines more the container than the content, focusing on the physical and architectural structure, the term "town," on the other hand, refers to the community, to the network of relationships and functions that include people, their activities, and their feelings of belonging and neighborliness. Proximity is an essential value, the foundation of social cohesion, a space in which to manage conflicts and build mutual recognition that influences the sphere of dignity and responsibility. In the town, individuals are never alone and are always intertwined with the community dimension. Individual behaviours become collective practices, and personal interest aligns with collective interest. This dynamic balance is what makes towns so unique and vital. The word "town" goes

Fig. 1.2 **A** Tamera, Portogallo © Italia che cambia, 2020; **B** Freiburg Vauban, Germania © Visit Freiburg, 2023; **C** Lavafante, Spagna © Corriere della sera, 2022; **D** Tänikon Smart Village, Svizzera © Agroscope, 2023

beyond a simple geographic description and represents, in fact, a deeper concept, based on community, human relationships, and mutual responsibility. Proximity, both geographic and social, is a fundamental value that contributes to social cohesion by creating a unique social fabric [29]. Unfortunately, the rhetoric of "small is beautiful" often reduces itself to superficial idealization. Towns are not just picturesque places for a romantic escape, summer holidays, or respite from the chaos of the cities. On the contrary, they are spaces of work, innovation, and great resilience. In rural communities, agriculture, livestock farming, and craftsmanship play crucial roles in the local economy and culture, often the result of traditions passed down through generations. Short supply chains, the collective management of common goods, and cooperative micro-enterprises represent sustainable economic models that can help keep these areas alive. However, to do justice to these communities, we must abandon the uniform approach to policy and services. The diversity of Italy's regions requires differentiated policies that consider local peculiarities and territorial vocation.

It is not possible for towns to be transformed into leisure spots for weekends or just tourist destinations; the typical "*quick getaway*" spots. Their authenticity and vitality depend on a constant commitment to preserving and promoting what makes them special. We need to rediscover and enhance the unique resources of Italy's inner areas, embracing a new way of living, producing, and consuming that goes beyond "*small-townism*". The way forward to save towns is clear: invest in agriculture, livestock farming, and craftsmanship, promote solidarity and cooperation, and adopt policies tailored to the specific needs of local communities, as well as focus on the most authentic architectural heritage, which is often better preserved but far from contemporary living standards. Only through this differentiated and inclusive approach can we protect and enhance towns, preserving their authenticity and vitality, and contribute to building a more sustainable and inclusive future for all.

The challenges faced by small communities in Italy are evident when considering the limitations imposed by their small size, a significant obstacle compared to the large-scale needs required by the modern economy. This situation often puts rural areas at a disadvantage in favour of larger, more organized centres that can provide higher-quality services. A disparity particularly evident in inner areas, especially along the Apennine belt, which, unlike the Alpine passes connecting to the northern states, have undergone less infrastructure development and, consequently, have been subject to progressive abandonment.

1.4.1 Minor Italian Villages

There are just over 300 villages, with a total population of approximately 6,000 inhabitants, certified by the Superintendent of Fine Arts as architectural and natural heritage sites. According to 2020 data presented by *Coldiretti*, 16% of Italians—around 9.8 million people—reside in these charming towns, which make up 54% of the national territory, reaching as much as 70% in some regions. In addition to those already internationally renowned, many other Italian villages have recently embarked on redevelopment and revitalization programs, not solely for tourism purposes. Stunning landscapes, the promotion of traditions, local food and wine production, and unspoiled natural environments characterize these villages, making them increasingly appealing not only to tourists but also to those seeking to purchase a second home or relocate permanently. Currently, villages represent a significant part of Italy's administrative structure, with 5,521 villages accounting for 69.85% of the country's 7,904 municipalities. Their geographical distribution indicates a higher concentration in the inner areas of the peninsula, along the Apennine ridge and the Alpine arc. Since the post-World War II period, inner areas have experienced gradual depopulation, losing 900,000 residents. The population of small municipalities recorded an average annual decline of 0.2% between 1981 and 2020. This decline coincided with migratory flows of young people and an increase in the average age of residents, with 17.6% of the population now over the age of 65. Culturally, villages represent a substantial part of Italy's heritage, encompassing 31.1% of sites and assets and 32.8% of museums. Their impact on food and wine culture is equally significant, with 90% of PDO (Protected Designation of Origin), PGI (Protected Geographical Indication), and *slow-food* products originating from small municipalities [30]. Associations such as *Borghi più belli d'Italia* (Most Beautiful Villages of Italy) and *Borghi Autentici d'Italia* (Authentic Villages of Italy) play a crucial role in identifying the most beautiful villages. Beyond creating networks and sharing enhancement projects, these associations work to promote these areas nationally and internationally, fostering tourism and local economic development while ensuring the preservation of artistic and cultural heritage.

Exploring these gems scattered throughout Italy is made possible through the associations' websites, categorized by region. Additionally, the television program *Borgo dei Borghi* (Village of Villages), broadcast by RAI, annually ranks Italy's most beautiful villages, allowing the public to vote for their favourite. Municipalities meeting the criteria can apply for the prestigious recognition of being named the "most beautiful village in Italy" [31] (Fig. 1.3). These picturesque settlements can be divided into three main categories: coastal, hillside, and mountain villages, each with its unique characteristics and distinct atmospheres. *Coastal villages* are located near the sea, enjoying a privileged position that offers not only breathtaking natural beauty but also direct access to marine resources. These villages often benefit from broader services and proximity to urban centres, making them attractive to visitors. However, some may find themselves in a state of neglect, lacking adequate administration to promote their development and enhancement. *Hillside villages*, by contrast, occupy an intermediate position between valleys and hilltops, often surrounded by vineyards and scenic landscapes. These villages may enjoy better accessibility compared to mountain ones, facilitating access to essential services. Their strategic location between elevations contributes to a relatively higher quality of life, although they may still face challenges related to managing local resources. Finally, *mountain villages* are Italy's hidden jewels, often perched on rugged peaks. While they face challenges such as limited accessibility and harsher climates, these villages retain a unique authenticity and charm. Their traditional architecture, often untouched by recent restoration efforts, bears witness to a history and culture that endure over time. Geographic isolation makes these towns less populated but, at the same time, helps preserve their intrinsic beauty and authenticity.

1.4.2 Abandonment and Depopulation

Addressing the issue of abandonment, small towns in Italy represent an alarming and widespread reality across the national territory, requiring serious attention. According to data provided by ISTAT, there are 5,627 small municipalities at risk, of which 3,532 have fewer than 2,000 inhabitants, and 2,105 fewer than 1,000 inhabitants, indicating an advanced state of depopulation. Among these, 197 villages are completely uninhabited. These communities, often consisting of clusters of houses surrounding a central square, typically the one near the main church or fountain, are in remote and hard-to-reach areas, frequently far from major communication routes. Most of these settlements are concentrated in the Apennines and inner regions [30].

More than mere clusters of houses, these villages are custodians of invaluable local identities. Through their way of life, methods of construction, and inhabitation

Fig. 1.3 **A** Fontainemore, Aosta Valley—Alpine Village © Borghi più belli d'Italia, 2024; **B** Campiglia Marittima, A Hymn with All the Shades on Nature's Musical Staff—Tuscany © Borghi più belli d'Italia, 2024; **C** Sepino, At the Crossroads of the Tratturi—Molise © Borghi più belli d'Italia, 2024; **D** Cisternino, L'abbagliante kasbah della valle d'Itria—Puglia © Borghi più belli d'Italia, 2024

patterns, they offer a glimpse into the past and the relationship between communities and the surrounding landscape. However, the abandonment of small towns in Italy has deep roots, dating back to the post-World War II period. During the economic boom, there was a massive exodus as people migrated to cities in search of jobs and prosperity, leading to the abandonment of agriculture in favour of industrialization. Adding to this analysis is the exodus of countless young people who, due to a lack of job opportunities, left southern Italy to settle in the north or abroad in search of better working conditions. Moreover, some villages were entirely abandoned because \schools were closed, like is underlined in the italian movie titled *Un mondo a parte* of Medusa Production, or due to natural disasters rendering them uninhabitable.

Calabria holds the record with 20 abandoned villages, followed by Tuscany and Piedmont, each with 19 abandoned settlements. *Abruzzo* and *Liguria* follow with 17 each, and *Sardinia* with 16. Notably, all regions, including those with strong economies, have at least one abandoned village. However, the Alps, with their robust economies, have not experienced depopulation to the same extent as the Apennines. *Valle d'Aosta*, *Trentino-Alto Adige*, *Veneto*, and *Friuli Venezia Giulia*, with a limited number of abandoned villages, clearly reflect this difference. The number of abandoned towns in Italy [30] exceeds a thousand, reaching about 6,000 when small residential clusters and alpine pastures are included.

Italy is witnessing a decline in its population. In 2018 alone, the Italian population decreased by 124,427, bringing the total population below 60 million. Without the contribution of foreign nationals, from 2011 to 2019, the Italian population would have decreased by over 800,000. The phenomenon of depopulation is thus an established fact. These trends, linked to an aging population and declining birth rates, became even more concerning in 2020 due to increased mortality caused by COVID-19 [30]. It is worth noting that such statistics are not uniformly distributed across the national territory. Demographic trends vary significantly: the population is growing in northern and central regions while declining in the south. Furthermore, there is population growth in urban and coastal areas, contrasted by decreases in small towns and inner areas. The National Institute of Statistics (ISTAT) provides a detailed picture of these dynamics through the Permanent Census of Population and Housing. Unlike the decennial censuses conducted before 2011, the current methodology is based on a significant annual sample, providing constantly updated information on various aspects of Italian life. Data from 2018 and 2019 published by ISTAT, including a specific focus on The Most Beautiful Villages in Italy, offer additional details. Of the 307 municipalities awarded this title, interesting and representative demographic trends emerge concerning depopulation dynamics, particularly in small municipalities in inner areas. Between 1951 and 2019, only 12 of these municipalities recorded consistent population growth, while 83 experienced continuous depopulation. The foreign population has proven crucial in mitigating demographic losses, with a significant increase after 2001. However, the analysis shows that villages with fewer than 10,000 inhabitants, especially those under 3,000, are the most affected, with a decrease of nearly 300,000 people since 1951. Despite an overall population loss in villages of approximately 185,000 individuals between 1951 and 2019, a reversal of this trend has been observed since 2001, thanks to immigration. This phenomenon intensified after 2011, particularly in medium-sized municipalities (5,000–15,000 inhabitants). Currently, the foreign population accounts for 7% of the population of the 307 most beautiful villages in Italy, a significant increase from 2.2% recorded in 2001. This foreign population is young and exhibits a higher fertility rate than the Italian average, suggesting a potential contribution to slowing depopulation in Italy. Strategies to counteract this phenomenon include initiatives to repopulate villages by leveraging their tourism potential and investing in digital infrastructure to enable remote work. The pandemic and ecological crisis may encourage a more sustainable lifestyle in villages. However, decisive political and economic commitment is essential to ensure the success of these initiatives. The National Strategy for Inner Areas (SNAI), launched in 2012, represents a significant example, albeit with some critical aspects, of an initiative to counter depopulation and revitalize territories. Its revival could significantly contribute to mitigating the phenomenon on a national scale.

1.4.3 A New Life for Minor Italian Villages

In Italy, the idea of creating *Smart Villages* in small towns, or rather, villages, is gaining increasing traction as an innovative solution to revitalize these rural areas, mitigate abandonment, and promote sustainable development while embracing modern digital technologies. In recent years, tourism focused on exploring Italian villages has seen significant numbers of visitors, generating growing interest and success.

Smart villages in Italy can be distinguished by several key features. First, they promote the adoption of high-speed connectivity, *IoT* (*Internet of Things*) solutions, and digital applications to improve services and stimulate innovation. Additionally, they place strong emphasis on sustainability by enhancing local resources, promoting energy efficiency, renewable energy use, sustainable land management, and eco-friendly agricultural practices. Finally, they are committed to preserving and enhancing cultural heritage, promoting the safeguarding of traditions, local productions, historical and artistic sites, and local architecture.

The benefits that *Smart Villages* bring to rural communities are manifold. First, they stimulate economic development by attracting new businesses and investments, promoting the sustainable growth of local activities through innovation and entrepreneurship. Moreover, the adoption of digital technologies and innovative solutions improves access to essential services, such as healthcare, education, transportation, and security, contributing to a better quality of life for residents. From an environmental perspective, they encourage sustainable practices and a transition toward a low-carbon economy by using renewable energy, efficiently managing resources, and promoting sustainable agriculture. Furthermore, they promote the preservation of rural communities' cultural identity by enhancing traditions, culture, and local heritage, and encouraging sustainable tourism and the promotion of local products. Finally, they aim to improve the accessibility of these areas. They represent an innovative model for rural development in Italy. The idea is that, soon, *Smart Villages* will become increasingly widespread in Italy, following the example of what has already happened in other European and non-European countries, but also drawing inspiration from several Italian cases already undertaken. Currently, there are various types of rehabilitation and enhancement projects for abandoned or at-risk villages, each with specific objectives, particularly:

- <u>tourist-real estate</u>: this is the most popular form of redevelopment as it ensures a quick return on investments. The focus is the transformation of existing buildings, even in a state of disrepair, into accommodation facilities or private residences. An example of this is the "*alberghi diffusi*" model, where the reception area is in a central building in the town and the rooms for hosting tourists are in surrounding houses, representing the most common and replicated model;
- <u>productive</u>: from ancient factories to modern start-ups, from university research centres to agricultural enterprises. An example is Solomeo, in the province of Perugia, which has been transformed into a centre for cashmere production;

- social: promotes local communities, develops innovative services, and builds supportive networks. The active involvement of various *stakeholders* is essential to foster the economic, social, and cultural growth of the community. Although there are no pure examples of *Social Housing*, forms of *Co-Housing* are being experimented with in places like the Torri Superiore village in Liguria;
- creative: focuses on art, culture, and entertainment. Activities range from the establishment of artistic communities to the organization of workshops, internships, artistic productions, and cultural events. Interaction with artists and cultural figures contributes to creating an organizational network and spaces to host events;
- work of art: uses creativity as a driving force for the development of smaller centres, enhancing the natural landscape and architectural heritage to attract new tourists in an alternative way.

In Italy, numerous successful practices have been implemented that can be considered early examples of "*Smart Villages*". However, to further promote the adoption of these development practices, it is essential to implement supportive policies and undertake additional initiatives involving multiple strategies. This can be achieved through dedicated funding, such as European, national, and regional funds for innovation and rural development projects, as well as through forms of collaboration between private entities, citizen initiatives, and partnerships with universities and research organizations. It is crucial to encourage collaboration between institutions to successfully implement *Smart Village* projects by combining expertise, resources, and knowledge. Actively involving the local community in the decision-making process and the management of *smart villages* fosters the sustainability of initiatives and creates a sense of belonging and active participation.

The Covid-19 pandemic has made it necessary to reconsider traditional development models, opening the way to new possibilities and solutions. This crisis highlighted the importance of a careful and gradual approach to transition, with a renewed awareness of the innovative challenges facing Italy's inner areas. The need to develop new ideas and development paradigms became more evident, bringing to light the problems caused by excessive urbanization and mobility, which contributed to the spread of the virus. This awareness has driven the consideration of social and urban decentralization solutions as a response to the current crisis, but also as an opportunity for secure and sustainable development. Examples include the first villages created to meet the need for *smart* working. Furthermore, the issue of depopulation in small municipalities and the desertification of decentralized areas has become even more pressing. In this context, *Smart Villages* present themselves as an innovative model for rural development; in 2023, there are many first smart examples across all Italian regions, the majority of which are in the mainland, with only a few, very few examples in Southern Italy and the islands. These examples feature a variety of projects and initiatives. The most common topics addressed include:

- improvement of energy efficiency in buildings: The use of advanced technologies to reduce energy consumption and promote sustainable energy sources;
- sustainable mobility: The adoption of eco-friendly transportation solutions such as electric vehicles or shared bicycles;

- smart waste management: The implementation of more efficient waste collection systems and recycling;
- security: The use of surveillance systems and other technologies to ensure community safety;
- promotion of sustainable agriculture: The use of advanced agricultural technologies to improve the productivity and sustainability of local farming;
- enhancement of cultural heritage: The preservation of cultural and historical traditions through tourism and cultural promotion initiatives.

Furthermore, they offer a range of benefits to rural communities, including:

- improvement of quality of life: Access to advanced and sustainable services, from healthcare to education, enhances residents' lives;
- promotion of sustainability: The adoption of renewable energy and eco-sustainable agricultural practices contributes to environmental protection;
- technology and innovation: The implementation of digital technologies stimulates economic and social innovation;
- attraction of new investments: *Smart Villages* become attractive for new investments, promoting economic growth;
- creation of new jobs: *Smart Village* projects often generate large, important, and essential local job opportunities.

On the global stage, as extensively explored in the previous sections, there is an increasing importance of services and the adoption of digital approaches across various sectors. However, in Italy, this transition is often slowed down by an architectural heritage of immense value and prestige. It is not, in fact, easy to introduce technologies and digitalisation into well-preserved contexts. Italy, with its rich history and unique architectural characteristics, faces challenges in the rapid implementation of new technologies. The slow adoption of technological systems can be attributed to the need to reconcile innovations with architectural contexts steeped in history and tradition, where the features are not only evident but often binding, especially regarding architectural heritage. Introducing digital solutions in environments characterized by ancient building structures can prove to be a complex, laborious, and extremely difficult process. The challenges arise not only at the level of technological adaptation but also due to the stringent regulations aimed at preserving the country's cultural heritage. The safeguarding of historic buildings, monuments, and archaeological sites imposes limitations and requires a careful, respectful, and contextually compatible approach in both the Alps and the Apennines [32]. As a result, while in the rest of the world the adoption of advanced technologies is often quick and seamless, in Italy there is a more measured balance between the heritage of the past and the needs of the present. On the one hand, this approach reflects the desire to preserve cultural heritage, but on the other hand, it can cause some delay in the adoption of new digital solutions, even when these could bring significant improvements in terms of efficiency and services. However, by using advanced technologies and the participatory approach of the community, it is possible to improve the quality of life in rural areas, stimulate economic growth, and promote sustainability [33–35].

The widespread adoption of these models will depend on the presence of supporting policies at the national and regional levels and the collaboration between public and private sectors to successfully implement the initiatives of *Smart Villages.*

1.5 European Regulatory Framework

The promotion and development of *Smart Villages* represent a priority of significance not only at the Italian level but also at the European and extra-European levels. The importance of these initiatives is emphasized by their connection to the European *Green Deal*, which, with its ambition for climate neutrality and ecological transition, offers a unique opportunity for territorial development [36]. The European *Green Deal* offers an ambitious and cross-cutting framework, which can be synergistically integrated with the development of *Smart Villages.* Key sectors such as energy efficiency, renewable energy sources, the circular economy, sustainable mobility, and sustainable agriculture represent areas where innovation can significantly contribute to the goals of the *Green Deal.* Funded projects such as *Smart Rural Communities*, *RURBAN+*, and *REScoop* illustrate the concrete commitment to the development of smart and sustainable communities in Europe. Integrating the *Green Deal* into the strategies and funding plans for *Smart Villages* is crucial to ensure a transition to a sustainable and resilient future in rural European areas [37].

At the European level, the European Commission has implemented several initiatives and funding programs, such as the European Strategy for *Smart Communities*, the Horizon Europe Program, the European Regional Development Fund (ERDF), and the European Agricultural Fund for Rural Development (EAFRD). These tools provide the necessary financial support to encourage the development of *Smart Villages* in a broader European context. In particular, the European Strategy for *Smart Communities*, which aims to help communities become smarter and more resilient through the use of digital technologies; the Horizon Europe Program, which funds research and innovation projects in fields such as artificial intelligence, robotics, sustainable mobility, and other key sectors for *Smart Villages* development; the *ERDF—European Regional Development Fund*, which funds economic and social development projects in rural areas, including those related to *Smart Villages* development; and the *EAFRD—European Agricultural Fund for Rural Development*, which funds sustainable rural development projects, including those related to *Smart Villages* development. In Italy, the PNRR (National Rehabilitation and Resilience Plan), the PSRN (National Rural Development Program), and the 2023 Budget Law are just some of the key tools in funding projects related to digitalisation, sustainability, and participation in Italian rural areas. The development of *Smart Villages* represents a priority at both the European and Italian levels. Since the case study analysed concerns an enclave within the European Union, all European strategies and regulations will be addressed and analysed, excluding other policies adopted outside the EU. The European Union has long recognized the potential of *Smart*

Villages to promote economic growth, social cohesion, and environmental sustainability in rural areas. In line with this goal, the European Commission has adopted a series of initiatives and funding programs to support the development of smaller territories. These initiatives and funding programs are contributing to creating an ecosystem conducive to the development of *Smart Villages* in Europe. Thanks to these interventions, an increasing number of rural villages are adopting innovative solutions to improve the quality of life for their residents, promote economic growth, and protect the environment. They represent a truly important opportunity for rural areas in Europe because, using digital technologies, they can improve the quality of life for residents, promote economic growth, and protect the environment. More European rural villages are becoming *smart*, and this trend is expected to continue growing in the coming years.

1.5.1 The Faro Convention

The Faro Convention, also known as STCE No. 199, is a crucial agreement that outlines and emphasizes key elements related to cultural heritage, especially in relation to human rights and democracy. It aims to promote a more inclusive vision of cultural heritage and its relationship with communities and society. Particularly significant is its reference to places of interest, which should be appreciated for the meanings and uses people attribute to them, as well as for the values they represent. Defined as a "framework convention," it outlines relevant themes, general objectives, and potential areas of intervention for member states. These states have the discretion to identify the most suitable means for its implementation, referring to their legal, institutional frameworks and specific experiences; it does not impose specific action obligations but suggests guiding principles.

Adopted by the Committee of Ministers of the Council of Europe on October 13, 2005, and open for signature by member states in Faro, Portugal, on October 27 of the same year, the Convention entered into force on June 1, 2011. Currently, twenty-five Council of Europe member states have ratified the Convention, while four states have signed it. The Convention is named after the Portuguese city of Faro, the venue for the inaugural event for the signature of Council of Europe member states on October 27, 2005. Italy's accession, which took place on February 27, 2013, in Strasbourg, brought the number of signatory states to 21, out of 47 Council of Europe members, with 14 of them having ratified the Convention's principles. As the latest of the international cultural conventions, it is based on the premise that the knowledge and use of cultural heritage are individual rights recognized in the Universal Declaration of Human Rights [38] and protected by the International Covenant on Economic, Social and Cultural Rights [39]. This approach enables individuals to freely participate in the cultural life of the community and enjoy the arts.

The Convention integrates with existing international instruments, inviting populations to play an active role in recognizing the values of cultural heritage and encouraging states to promote a participatory valorisation process involving public institutions, private citizens, and associations. It aims to harmonize European valorisation policies, considering ongoing processes of democratization of culture and open government. It recognizes that the active participation of citizens and communities is key to increasing awareness of the value of cultural heritage in Europe and its contribution to well-being and quality of life. It represents one of the approaches to cultural heritage that aligns with the concept of *Smart Villages.*

The choice of Fontecchio as the venue for the first research-action workshop on the Faro Convention, a small village with about 400 inhabitants located in the *Aterno* Valley in *Abruzzo,* is not incidental. In 2013, Fontecchio became the first small Italian municipality to adhere to the principles of the Convention, highlighting a deeply rooted sensitivity within the community towards its cultural heritage, both tangible, intangible, and natural. During the 2009 *L'Aquila* earthquake, Fontecchio had already demonstrated a strong awareness of the importance of cultural heritage and the artistic and architectural assets connected to it, which were considered by the community as strategic and essential elements for passing on memory to future generations. Immediately after the tragic event, the "*Borghi Attivi*" project was launched, promoting a deliberative democracy process to outline guidelines for local development and the village's aesthetics, also aiming to involve private citizens with the main goal of providing housing with good or excellent quality standards at subsidized rents. From these guidelines, the "*Casa&Bottega*" initiative emerged, a collective tool to defend employment and thc landscape, supporting social housing and urban regeneration.

1.5.2 The Cork 2.0 Conference

The Cork 2.0 Conference, held on September 5–6, 2012, in the city of Cork, Ireland, was an important event promoted by the European Commission in close collaboration with the Irish government. This conference brought together over 300 participants from all EU member states and various countries around the world, with the aim of addressing rural development issues in Europe. It marked a crucial milestone in the evolution of European policies related to the development of rural regions. During the event, participants had the opportunity to engage in a thorough debate on the challenges and opportunities facing Europe's rural areas and, at the same time, develop strategies and solutions to successfully address them. A significant outcome of the Cork 2.0 Conference was the adoption of a new document updating the 1996 Cork Declaration. This document placed particular emphasis on the crucial role of rural areas in Europe's sustainable development, while identifying a series of goals and priorities essential for future rural development initiatives. Among the key objectives outlined by the Cork 2.0 Declaration were the enhancement of quality of life in rural regions, with particular attention to the provision of essential services and access to educational and training opportunities; the promotion of economic growth and employment within rural communities, recognizing their undeniable potential for innovation and entrepreneurship; and the preservation of the environment and natural resources in rural areas through the encouragement of sustainable farming practices and the implementation of measures to protect the land, inside the Cork Declaration of 2016). At the same time, the Cork 2.0 Conference marked the beginning of the European Union's rural development program for the 2014–2020 period, with a financial allocation of over 100 billion euros to support projects aimed at rural development. The legacy of the Cork 2.0 Conference is still tangible today, with numerous EU member states having implemented a series of projects and policies to promote sustainable rural development. The conference strengthened the central role of rural regions within the European context and laid out a clear vision for the sustainable development of rural communities in Europe. The influence of this event endures, having a lasting impact on rural development across continental Europe.

1.5.3 EU Action Plan for Smart Villages

On April 11, 2017, the European Commission presented the EU Action Plan for *Smart Villages*, which incorporates some conclusions from the 2016 Cork 2.0 Conference on "A Better Life in Rural Areas." This plan is based on the potential and development of new opportunities through technological innovation and environmental sustainability. The goal of the European Action Plan [40] is to promote sustainability and improve the quality of life in rural and mountainous areas. To achieve this, it aims to coordinate initiatives related to European policies such as rural development, regional development, research, transport, energy, and digitalisation. Specifically, its three main objectives are:

- digitalisation and connectivity: Promote the adoption of digital technologies and improve connectivity in rural areas, facilitating access to advanced digital services and opportunities;
- sustainability and green transition: Foster the transition to a sustainable economy by promoting efficient resource management, the use of renewable energy, environmental protection, and the adoption of sustainable agricultural practices;
- social inclusion and quality of life: Promote social inclusion, improve the quality of life in rural areas, and ensure access to essential services for all, such as education, healthcare, transportation, and culture.

The areas of focus of the Action Plan include:

- digital connectivity: The Action Plan commits to ensuring universal access to high-quality digital connectivity for all rural areas;
- technological innovation: Technological innovation is encouraged in rural areas, with particular attention to sectors such as agriculture, energy, transport, and social services;

- environmental sustainability: The Action Plan promotes environmental sustainability in rural areas through the adoption of digital technologies aimed at reducing environmental impact;
- local community participation: Actively supports the participation of local communities in the development process of *Smart Villages*;
- cooperation among stakeholders: The Action Plan encourages collaboration between public and private entities for the progress of *Smart Villages*.

The funding for the Action Plan was supported by an investment of 100 million euros within the framework of the European Union's Rural Development Program for the 2014–2020 period. Additionally, the European Commission launched a series of initiatives aimed at promoting the development of *smart villages*, including the "*Smart Villages*" portal. Tangible examples of how digital technologies can improve rural areas include:

- precision agriculture: The use of artificial intelligence technologies to optimize the use of agricultural resources, increasing productivity and reducing environmental impact;
- renewable energy: The management and monitoring of renewable energy resources through digital technologies to improve efficiency and accessibility;
- sustainable mobility: Improvement of public and private transport services in rural areas using digital technologies to facilitate movement;
- social services: Provision of healthcare, educational, and social assistance services remotely using digital technologies, increasing accessibility for residents of rural areas.

It represents a significant initiative for the progress of rural areas in Europe. It offers a wide range of resources and tools aimed at enhancing the use of digital technologies to improve the quality of life, foster economic growth, and preserve the environment within rural villages. Since these rural areas have specific needs and potentials, and lack standardized models or solutions, a development approach centred on territorial capital has been chosen. This refers to the set of social, environmental, economic, cultural, and institutional resources that characterize places. Thus, "*smart*" villages are born, which are communities made up of citizens who, through active participation, find concrete solutions to daily challenges, interpreting in innovative ways the dynamics that are transforming rural areas. The concept of "*smart*" is closely related to the idea of "*smart*" in terms of using digital technologies to promote social innovation, raise awareness of renewable energy and emerging bioeconomy, meet the growing demand for typical and healthy products, adopt circular economy principles, and promote ecotourism. Additionally, the creation of new forms of cooperation and alliances between the agricultural sector and various components of rural society, between municipalities, private entities, and civil society, is encouraged.

Numerous *Smart Village* projects have already been launched. EU initiatives aim to provide data, *best practices*, networks, funding, and regulations to support these experiences [40]. Some of these initiatives include:

- the *SIMRA Project*—Social Innovation in Marginalized Rural Areas, funded by the Horizon 2020 research and innovation program, defines a database for social innovation in the rural areas of the Mediterranean Basin, identifying *best practices* and their application in a pilot context;
- the *SMARTA Project*—Rethinking Rural Mobility, from the European Parliament, which proposes innovative solutions for rural mobility;
- the TESS Project—Transactional Environmental Support System, funded by the EU's 7th Framework Programme, which gathers 63 climate initiatives implemented by different communities across Europe, highlighting the climate change mitigation potential and social cohesion resulting from collaboration;
- the *RURITAGE Project*, coordinated by the University of Bologna, to promote local cultural and natural heritage as a catalyst for sustainable development, with six innovation areas to transform rural areas into sustainable development labs/demonstrators;
- the integrated *RurAction research and doctoral program*, funded by the European Union under the Marie Sklodowska-Curie Horizon 2020 Innovative Training Network actions;
- the Alpine Macroregional Strategy, officially *EUSALP* (EU Strategy for the Alpine Region), an agreement signed in 2013 by EU countries: Italy, France, Germany, Austria, Slovenia, and two non-EU countries, Switzerland and Liechtenstein, to promote the sustainable management of energy, natural and cultural resources, and environmental protection of the territory, aiming to enhance the development of the Alpine area;
- *RUMRA and Smart Villages*, the Intergroup on rural, mountain, and remote areas (2014–2019), promoting integrated development for European territories in their diversity, supporting new forms of *governance* and cooperation, place-based initiatives, and the use of technology, with the goal of strengthening integration and ensuring a sustainable future for all Europeans;
- the *SOCLIMPACT Project*, which aims to mitigate the effects of climate change and their socio-economic impact on European islands from 2030 to 2100 in the context of the Blue Economy, exploring and evaluating the impact of potential decarbonization pathways;
- *COASTAL*, with the aim of promoting multi-actor collaboration among coastal and rural *stakeholders*, formulating and evaluating business solutions and policy recommendations to improve the cultural, social, and economic development of islands;
- *DESIRA* (Digitisation: Economic and Social Impacts in Rural Areas), a Horizon 2020 project (2019–2023) coordinated by the University of Pisa, involving 25 partner organizations (research entities, NGOs, and SMEs) in a multi-actor and interdisciplinary approach. The project aims to enhance society's and policy-makers' ability to respond to the challenges posed by digitalisation in agriculture, forestry, and rural areas;
- the *dRURAL Project*, which develops and implements a digital solution offering multiple services to rural citizens, creating opportunities, economic growth, and improvements in quality of life, funded by Horizon 2020;

- *RUBIZMO*, a 42 month project funded by the EU's Horizon 2020 research and innovation program, launched in May 2018, aiming to discover the vital ingredients for the development of entrepreneurship and successful business models in high-potential sectors such as food and agriculture, new bio-based value chains, and ecosystem services.

The EU Action Plan for *Smart Villages* represents a significant historical moment, introducing, for the first time, the theme of artificial intelligence and, therefore, the concept of "*smart*", as well as an operational tool to promote innovation, sustainability, and inclusion in rural areas. Through its guidelines and initiatives, the plan aims to transform rural communities by improving access to technology, promoting environmental sustainability, and fostering social inclusion. Successfully implementing the Action Plan requires ongoing commitment from European institutions, national governments, and local communities to build modern, sustainable, and prosperous *Smart Villages.*

1.5.4 The Smart Villages Action Plan

On May 22, 2018, in Brussels, the European Rural Network presented the action plan for *Smart Villages*, with a strategy based on 16 initiatives to support and improve the implementation of EU policies aimed at rural areas. The goal is to reflect on the villages of the future, announcing a series of initiatives within the framework of EU policies and funds related to rural development, regional development, research, transportation, energy, and digitalisation. The action plan includes 16 initiatives, including conferences, seminars, and pilot projects, ongoing or scheduled until 2020. These are measures already activated and funded across various intervention lines, aimed at improving the implementation of EU policies for rural areas and new initiatives directly linked to *Smart Villages* [41].

The European Rural Development Network (RESR) has highlighted that for many European citizens, rural areas are a place to live, work, and raise a family. The EU's actions align with this vision and exemplify how the existence of common institutions makes it possible to spread good practices to address similar challenges, with institutional incentives toward multi-level *governance.* Such *governance* strives to create convergence among the many tools available to different institutions (local, regional, national, European), ensuring a balance between the need for uniformity in rights and duties and the recognition of the diversities present in communities and territories.

The goal of the involved Directorates-General is to identify drivers and success factors for *smart villages*, that is, rural communities that fully leverage their resources, enhancing traditional ones and, most importantly, activating new ones through digital technologies, communication, and innovation to create opportunities and attractiveness for rural areas. The project's outcome should lead to identifying solutions applicable across the European Union [42], considering the diversity of different contexts.

Sharing a common definition and characteristics of eco-social and *smart villages*, with concrete examples of good practices, should be an important tool to contribute to improving rural development. The involvement of all citizens supported by effective *governance* will ensure success.

There are numerous European policies contributing to the realization of this project, including rural development policies, cohesion policies, and calls under programs like Horizon 2020 and Connecting Europe Facility, through a vision of collaboration and integration of available EU funds. However, it is primarily the European Agricultural Fund for Rural Development (EAFRD) (also with LAEDER and PEI) that plays a central role, providing numerous support tools for *smart villages*. Nearly 100 billion euros from the EU budget have been allocated for a total of 118 rural development projects for programs in the period 2014–2020. These investments primarily concern the environment, climate, and transport sectors.

These are the 16 initiatives of the Action Plan:

- TWG "Smart Rural Enterprises"—European Rural Network—2016–2017;
- TWG "*Smart Villages*"—European Rural Network—second half of 2017;
- Workshop on access to data and digital platforms—PEI—April 2017;
- Seminars on developing new projects using technologies—PEI—June 2017;
- Conference on innovation in agriculture and rural development—Portugal and DG AGRI—October 2017;
- Call for pilot projects on Smart Eco Villages—DG AGRI—2017–2018;
- LEADER/CLLD—Member States—2014–2020;
- Smart specialization on agri-food (S3P Agri-Food)—DG REGIO, AGRI, JRC, RTD—2014–2020;
- Broadband Competence Network (BCO)—DG AGRI, REGIO, CNECT—2016–2020;
- Digital *Innovation Hubs* (DIH)—DG CNECT and RTD—2014–2020;
- SMARTA—DG MOVE—since 2016;
- Internet of Things (*IoT*)—Horizon 2020—DG AGRI and CNECT—2017–2020;
- More efficient and smarter door-to-door logistics—Horizon 2020—DG MOVE—2016–2017;
- Business models for a modern rural economy—Horizon 2020—DG AGRI—2016/2017;
- Future calls under "rural renaissance"—Horizon 2020—DG AGRI—2018–2020;
- Information-based bioeconomy—Horizon 2020—DG CNECT—2017–2019.

In the last two years, the European Rural Development Network (RESR) has carried out various activities to examine the first of these two challenges, organizing thematic groups on value chains and businesses and their smart competitiveness. The group aims to become a reference point for sharing experiences and promoting new ways of implementing rural development policy [43]. The main objectives are therefore: to develop recommendations and examples to improve strategies for revitalizing innovative rural services and strengthening the role of RDPs, to provide guidance, tools, and examples at the level of concrete projects. All *feedback* from these activities will then be disseminated through the rural development network. In the Italian context, many small municipalities and rural villages can embark on the *Smart Village* path and can serve as examples of sustainability and opportunity. The National Strategy for Inner Areas, which we will discuss later, focuses on these rural realities to guide them, leveraging their strengths, consolidating them, and creating new ones to offer new development opportunities [44].

1.5.5 Next Generation EU

In recent years, the European Union has gone through a period of repeated financial crises and recessions. Income and employment inequalities between member states and across geographic areas have increased, exacerbating social tensions and political risks. In 2020, the health emergency caused by the Covid-19 pandemic led to an unprecedented production shutdown and the adoption of emergency measures with profound social consequences. During the Covid-19 epidemic, the awareness of common fragility was accompanied by the urgency for a turning point. Europeans approved tools aimed at addressing the crisis and laying the foundations for rehabilitation. The real historic turning point came with the agreement to finance the *Next Generation EU* (NGEU) initiative with €750 billion. All institutions were involved in a coordinated and consistent action for this new course. Already in the first half of 2020, the European Central Bank (ECB) continued and strengthened its monetary expansion. The Commission approved the flexible use of budget resources through the Coronavirus Response Investment Initiative (CRII plus), the general safeguard clause of the Stability and Growth Pact, and the temporary framework for state aid rules. Even before NGEU, unprecedented safety nets were made available to address the social, economic, and health emergency: SURE, the recapitalization of the EIB, as well as a new specific line of credit to tackle the pandemic within the European Stability Mechanism (ESM), approved in spring and available to Euro Area member states (European Commission, 2021). Finally, at the end of December 2020, the Council of the European Union and the European Parliament reached an agreement on the new Multiannual Financial Framework for 2021–2027, overcoming vetoes in an effort of solidarity and making all available resources operational for the Union's recovery. Next Generation EU (European Commission, 2021) addresses a historic need, the creation of new intra-European and Mediterranean networks, and the launch of collaborative infrastructures for science and research, starting with the challenge of global health. It expresses the urgency and opportunity of a true European economic Renaissance, with an instrument that will allow the Commission to raise funds on the capital markets. It signals the sharing of a common course. A cultural challenge that involves all territories. A transformation process toward *green* and digital transition, enabling the European Union to regain ground in the global technological race, create quality jobs while maintaining and renewing the European social model, and establish global leadership for sustainable development, more necessary after the Paris Climate Agreements.

Italy has been strongly committed to this European turning point. Today, it is on a path of progressively shared risks for investments aimed at addressing common priorities, recovering production capacity, improving both material and immaterial infrastructure, and tackling the energy and digital transitions. The challenge of inclusive growth concerns all of Europe, which must find a new role in technological competition and the reorganization of value chains. But it particularly concerns Italy, where previous crises have exacerbated significant gender, generational, and territorial inequalities, deeply undermining rehabilitation capacities. Italy intends to be a key player in this European Renaissance through the revitalization of public and private investments, alongside reforms aimed at strengthening the capacity and efficiency of institutions through the National Rehabilitation and Resilience Plan, which will be addressed in the following sections. This plan is focused on programming and implementing investments that mark a decisive break for sustainable development, digitalisation, and innovation, as well as reducing disparities and inequalities, particularly between rural areas and *Smart Cities*. The tool is designed to shape a more sustainable and inclusive Italy, equipped with an advanced and dynamic economy. As a recovery-oriented initiative, the Plan aims to address the economic and social impact of the pandemic crisis, drawing on lessons learned during some of the most difficult periods in Italy's republican history. Rehabilitation must shape a new Italy, leveraging the opportunities offered by the ecological and digital transitions. The goal is to unleash the growth potential of the economy, increase productivity, generate new employment, and improve the quality of work and public services, starting with healthcare and education. This resilience work involves preparing the state, businesses, and all social actors to face such events, adapting to the changes brought by globalization and new technological frontiers. Furthermore, it assumes a reformist character, as the investment lines are accompanied by the adoption of a reform strategy in line with the European Commission's Recommendations and National Reform Plans. The implementation of reforms is an integral part of the Plan.

The revitalization action outlined in the Plan is guided by policy objectives and interventions linked to the three European strategic pillars: digitalisation and innovation; ecological transition and social inclusion. These priorities play a crucial role for Italy, addressing critical issues such as the low propensity for innovation and the low level of digitalisation, which contribute to weak economic growth rates and insufficient employment. The Plan allows Italy to become a key player in the European Green Deal, reducing pollutant emissions, creating jobs in the *green* economy, improving energy efficiency, and promoting industrial processes related to the *green* transition. The challenge of sustainability and emission reduction will also be addressed through digital solutions.

In the next decade, rapid digital transformation will be central to geopolitical competition and will characterize the industrial supply chains of Italian manufacturing and all aspects of social life. Digital technology will be the enabling platform for reforms and competitiveness. Extensive investment in digital infrastructure, skills, and culture will unlock the potential of all Italian territories.

For Italy to be truly sustainable, it is necessary to address and reduce gender, generational, and territorial inequalities. Therefore, implementing the interventions connected to the Plan becomes essential to tackle and resolve issues related to three cross-cutting priorities: women, youth, and the South, which are crucial to activate the development potential and rethink social infrastructure and the public sector. The main mission is to align and govern the growth drivers of the next decade by involving all citizens and economic and social entities in a collective national effort, addressing the structural issues that have hindered Italian development for too long.

1.5.6 Youth4Climate, Pre-COP26, and COP26

Between September 28 and October 2, 2021, the Youth4Climate initiative took place, globally recognized in 2002 thanks to the collaboration between the Ministry of Environment and Energy Security and the United Nations Development Programme. This significant event involved young climate leaders in preparation for Pre-COP26. Aware of the urgent need to address climate change, they shared solutions and actively participated in the debate to positively influence the agenda. It is important to highlight that the initiative not only focused on global climate issues but also on the importance of extending its impact to Europe's inner areas. The event was promoted by the Italian Ministry for Ecological Transition, in collaboration with Connect4Climate of the World Bank Group, with the participation of the Lombardy Region and the Municipality of Milan, as well as the active support of *FAI—Fondo Ambiente Italiano* volunteers, Santagata Foundation, Italia Nostra, and *AIGU—Associazione Italiana Giovani* per l'UNESCO. The initiative brought together young global citizens to examine territorial issues, including both cities and peripheral centres, such as rural and remote areas in Italy. The discussions among the young leaders focused on four specific topics: the role of civil societies in adapting to climate change, the importance of ecological transition in cultural heritage, actions for safeguarding, emergency response, and prevention of fragile cultural heritage, and the use of computerized systems to understand risks related to cultural heritage. This meeting facilitated a fruitful exchange of ideas between experts and participants, culminating in the drafting of a document sent to the Ministry of Cultural Heritage and Activities (MIC) to integrate it into preparatory materials for COP26. On September 30, 2021, the youth's requests were heard by President Sergio Mattarella, Prime Minister Mario Draghi, and UK Prime Minister Boris Johnson during the opening of the Pre-COP26 ministerial conference. The Pre-COP26, which involved around 400 young delegates from various countries, was a crucial event to fuel the October Pre-COP summit. The young participants were called to develop four key pillars: Youth Driving Ambition, A Climate-Conscious Society, Sustainable Recovery, and Engagement. Finally, COP26, which concluded in Glasgow on November 13, saw significant progress, including an increase in financial commitments to support developing countries and the adoption of key initiatives such as reducing methane emissions and the development of the Paris Code. The Youth4Climate initiative and the active participation of young climate leaders represented hope and an opportunity for change, extending their contribution to Europe's inner areas. Their commitment was crucial in building a more resilient world and mitigating the effects of climate change, uniting governments, international organizations, civil society, and young leaders in building a sustainable future.

1.6 Italian Regulatory Framework

The commitment of the Italian government to the revitalization of the inner areas goes beyond mere fund allocation; it is a statement of intent aimed at renewing and revitalizing rural areas, often neglected. The PNRR emerges as a guiding beacon, offering a holistic vision that not only aims to counter depopulation but also to create conditions for sustainable economic growth and a better quality of life in the inner regions. The funding of 1.8 billion euros allocated to the inner areas of Italy within the PNRR highlights the importance of modernizing and digitizing these regions. The installation of high-speed Wi-Fi networks is not only a step towards global connectivity but also a powerful means to encourage innovation, training, and the active participation of rural communities in the digital economy. The adoption of sustainable mobility systems, promoted through this plan, will improve connectivity between villages, while also stimulating local and sustainable tourism.

The PSRN—National Rural Development Plan, with its funding of 1.5 billion euros, is an important complement, addressing key aspects such as agricultural innovation, job creation, and tourism attraction. The creation of new agricultural enterprises represents an investment in the future of rural communities, contributing not only to food security but also to economic growth. Attention to the transformation and commercialization of agricultural products enhances the role of these areas on the national stage, promoting their unique identity and stimulating the local economy. The 2023 Budget Law, with its additional funding of 100 million euros for *Smart Villages*, is another step forward. The installation of solar panels and photovoltaic systems in public buildings not only promotes environmental sustainability but also creates new job opportunities and reduces dependence on traditional energy sources. The promotion of sustainable mobility and the creation of training and innovation centres offer fertile ground for the development of local skills and initiatives. Observing the European context and inner areas that have already benefited from similar funds provides a valuable perspective for Italy. The diversity of approaches and the success of such strategies show how a long-term commitment and an integrated vision are essential for the revival of inner areas. Italy is adopting an ambitious and comprehensive strategy to ensure that its inner areas not only overcome the challenges of depopulation but become centres of innovation, sustainability, and prosperity. These investments represent an investment in the country's future, preserving and enhancing the cultural and natural heritage of the inner regions. The path taken seems to promise a future in which rural communities not only survive but thrive, actively contributing to the overall growth of Italy.

In the following paragraphs, the various initiatives aimed at supporting the development, rehabilitation, and enhancement of Italy's inner areas will be closely examined, drawing on European paradigms. First, the pillar of the National Strategy for Inner Areas (SNAI) will be examined, as it constitutes the foundation of the entire thematic context. Subsequently, the National Rehabilitation and Resilience Plan (PNRR), which in response to the pandemic event, has taken a central role in defining strategies for the relaunch and development of these regions.

In addition to these central programs, the analysis will cover a series of more specific interventions to promote the revitalization of the most picturesque places in our country. In this context, the National Plan for the Redevelopment of Small Municipalities will be explored, a key initiative aimed at revitalizing small communities while preserving their cultural and historical heritage. The National Strategy for Mountain Areas (SNAMI), which was born with the aim of supporting mountain territories, often disadvantaged, to ensure their economic and environmental sustainability.

The National Strategy for Climate Change Adaptation (SNAC), which aims to adopt effective strategies in response to the rapidly evolving climate challenges. Finally, the Action and Cohesion Plan (PAC) will be examined, a significant program aimed at ensuring equitable and sustainable cohesion between the country's regions and the regional initiatives undertaken for small rural communities. Taken together, these initiatives outline a complex but well-articulated framework of targeted interventions, all aimed at supporting the regeneration of Italy's inner areas. Through the analysis of these programs, the commitment of the Italian government to promoting sustainable development, resilience, and the enhancement of the wealth found in these fascinating regions can be fully understood.

1.6.1 SNAI—National Strategy for Inner Areas

The National Strategy for Inner Areas (SNAI) is a territorial development and cohesion policy adopted at the national level, aimed at combating the phenomenon of marginalization and depopulation in the inner regions of our country. This ambitious project introduced new forms of *governance*, adopting an integrated approach at the local level, to promote the development of marginal areas characterized by geographical and demographic disadvantages. The main goal of the SNAI is to address these problems and find solutions by combating demographic decline and halting the "*demographic haemorrhage*" that affects these areas. The SNAI selected 72 Italian areas, 1,077 municipalities, and about 2,072,718 inhabitants. For the 2014–2020 programming period, this included 1,060 municipalities with around 2 million inhabitants. These areas cover an area of about 51,000 km and represent 13.4% of all Italian municipalities and 26% of municipalities classified as inner areas. They also constitute 3.4% of the national population and 15.5% of the population residing in municipalities classified as inner areas. They occupy 17% of the entire national territory and 28.4% of the total area of all inner areas in Italy. National resources allocated to these areas amount to over 591 million euros, in addition to funds from the European Structural and Investment Funds (SIE) and other public and private funds to counter depopulation in inner areas and promote social cohesion.

The implementation actions focus on two main areas within the municipalities of the Inner Areas: essential services and local development. Regarding essential services, projects aim to improve the quality and quantity of education, healthcare, and mobility services. Local development actions, on the other hand, focus on protecting the territory and local communities, enhancing natural and cultural resources, supporting agro-food systems and local development, energy savings, and local renewable energy and craftsmanship supply chains. The SNAI also includes a series of actions to promote collaboration among the various inner areas, including the creation of the Network of Inner Areas, a cooperation platform between the different Italian inner areas aimed at promoting the sharing of experiences and collaboration between areas, to strengthen their development capacities. Furthermore, a series of pilot projects have been funded to test new solutions for the development of inner areas. The Network of Inner Areas has partially contributed to collaboration between inner areas and the strengthening of their development capacities, but there is still much to be done.

1.6.2 PNRR—National Rehabilitation and Resilience Plan

The pandemic struck Italy, exacerbating an economic, social, and environmental crisis already underway. Even before COVID-19, the country lagged other advanced European nations, grappling with productivity issues, a fragmented productive fabric, and a limited inclination to adopt new technologies, both in the private sector and in Public Administration. The Next Generation EU (NGEU), better known in Italy as the "Rehabilitation Fund," allocated funds to help Member States recover from the pandemic-induced crisis. To access these funds, Italy, like every other Member State, had to present its package of investments and reforms, known as the National Rehabilitation and Resilience Plan (PNRR). This plan is structured into six missions, with a budget of €191.5 billion allocated as follows:

- digitalisation, innovation, competitiveness, and culture: €40.32 billion;
- *green* revolution and ecological transition: €59.47 billion;
- infrastructure for sustainable mobility: €25.40 billion;
- education and research: €30.88 billion;
- inclusion and cohesion: €19.81 billion;
- health: €15.63 billion.

<u>Digitalisation, Innovation, Competitiveness, and Culture</u>: promoting digital innovation to support the country's infrastructure and the transformation of production processes in businesses, with a particular focus on Transition 4.0 and innovative projects for the Made in Italy supply chains. These projects are accompanied by a reform on industrial property aimed at defining a multi-year strategy to foster a

culture of innovation and the tools for protecting and enhancing industrial property. The reform, initiated by Minister Giorgetti, allocated €30 million in resources.

Green Revolution and Ecological Transition: promoting the country's *green* transition by focusing on energy production from renewable sources, increasing resilience to climate change, supporting research and innovation, and incentivizing sustainable public transport. Of these funds, €1.25 billion has been allocated to strengthen investments in key sectors of the ecological transition, while €1 billion has been dedicated to investments in renewable energy and batteries for the photovoltaic, wind, and battery industries. Key actions include the creation of a Gigafactory to produce high-efficiency innovative photovoltaic panels, the construction of an industrial facility for manufacturing flexible wind energy panels, and the development of an "ultra-modern" *Gigafactory* in the battery sector. Additionally, an investment of €250 million has been planned to support the growth of innovative *startups* through the creation of a *Green Transition Fund* (GTF) targeting sectors such as renewable energy, the circular economy, mobility, energy efficiency, waste management, and energy storage.

Education and Research: Research and development aim to enhance interaction between businesses and public entities, improving small and medium-sized enterprises' capacity for innovation and their participation in strategic sectors critical for national and European competitiveness.

The PNRR and complementary funds provide €3.35 billion, distributed as follows:

- funding for the IPCEI Fund: €1.5 billion;
- horizon Europe: €200 million;
- funding for the National Innovation Fund: €300 million;
- refinancing and thematic extension of technology transfer centres: €350 million.

In addition, €1 billion is allocated to finance the *Innovation Agreements.*

Inclusion and cohesion: With a funding of €400 million, the plan also includes strengthening and launching new female entrepreneurship. Additionally, the "Fondo Impresa Donna" (Women's Business Fund) was enhanced with the 2021 Budget Law, which allocated an initial funding of €40 million. Approximately €66 billion are designated for investments managed by territorial authorities, with €20 billion allocated to the South, plus €9 billion from the React-EU program, €54 billion from European Structural Funds, and €58 billion from the Development and Cohesion Fund. To support the actions of municipalities, provinces, metropolitan cities, and regions, the government has implemented a series of tools, especially for the South and local administration bodies. Specific technical figures are planned to aid at various stages of program development and project implementation. The available tools include:

- specific hires/1: municipalities will have the opportunity to hire non-managerial personnel with a dedicated fund of €30 million per year until 2026, aimed at reducing expenses for municipalities with fewer than 5,000 inhabitants;

- specific hires/2: additional opportunities to hire permanent staff for the implementation of PNRR projects;

SOUTH competition: recruitment of 2,022 additional units to support administrations in the South for the implementation of national and European cohesion policies, with more favourable criteria for young graduates;

- 1,000 positions for administrative activities: regions can establish 1,000 collaborative positions with professionals and experts to support administrative processes related to the implementation of the PNRR, with 40% reserved for the South;
- new forces in the South: collaboration contracts with specialized personnel and professionals;
- resources from European Structural Funds: possibility to use resources from the Complementary Operational Programs (POC) for technical and operational support;
- consulting for experts in large cities: municipalities with over 250,000 inhabitants can enter into consulting and collaboration agreements with qualified experts to accelerate projects;
- support for large municipalities in "pre-crisis": municipalities with over 250,000 inhabitants can strengthen the staff of offices under the mayor's administration;
- fund for idea and project competitions: municipalities with up to 30,000 inhabitants can organize idea and project competitions in the fields of urban planning and social innovation or use such resources for technical and economic feasibility studies;
- technical support from state-owned companies: qualified experts for technical-operational support from *Consip, Invitalia, Difesa Servizi*, and *Sogei*;
- support services for contracting authorities: staff training by Consip;
- agreement with *Cassa Depositi e Prestiti*;
- task force for school buildings.

Health: investments to strengthen the healthcare system, improve prevention, manage healthcare emergencies, and promote innovation in the health sector. This includes projects to enhance healthcare infrastructure, digitalize healthcare services, improve the training of healthcare personnel, and promote scientific research in the field of health.

To finance additional interventions, the Italian government has also approved an additional fund of €30.6 billion, bringing the total to €222.1 billion. The Ministry of Economic Development (MISE) is responsible for ten investment projects and one reform project. Each project has been designed to follow the "Specific Recommendations" of the European Commission while strengthening the country's growth potential. The PNRR proposal focuses on three key areas of intervention shared at the European level: digitalisation and innovation, ecological transition, and social inclusion. According to the estimates in the same plan, the Government expects a 3.6 percentage point increase in GDP and a 3.2 percentage point increase in employment by 2026. The plan is structured into four areas: General objectives; Reforms and Missions; Implementation and monitoring and Macroeconomic impact assessment.

Going into more detail, the PNRR is structured along 16 Components, aimed at achieving the economic and social objectives defined in the Government's strategy. These components are grouped into 6 Missions, for each of which the necessary reforms for more effective implementation of interventions are outlined. To prepare for the implementation of the PNRR, the Ministry of Culture (MiC in Italian) has internally organized the establishment of a mission unit at the managerial level within the central administrations. This unit is responsible for communicating with the Coordination Unit of the Presidency of the Council and the Central Governance Unit of the Plan at the Ministry of Economy and Finance (MEF). Additionally, in the field of protection and enhancement of built heritage, there is a Special Superintendency for the PNRR, which will be responsible for the protection of cultural and landscape assets affected by the rehabilitation plan's interventions, subject to Environmental Impact Assessment (VIA) at the national level or falling within the territory of at least two peripheral offices of the Ministry (DPCM 6/2021). This Superintendency, referred to as the General Directorate, is entrusted to the General Director for Archaeology, Fine Arts, and Landscape (ABAP) of the Ministry of Culture (MiC) and is supported by a Technical Secretariat. Sixty-three reforms in total, which can be divided into the following types:

- horizontal reforms (or contextual reforms), structural innovations in the legal system aimed at improving equity, efficiency, competitiveness, and the economic framework of the country;
- enabling reforms to ensure the implementation of the Plan and, in general, to remove administrative, regulatory, and procedural obstacles that affect economic activities and the quality of services provided;
- sectoral reforms (included in the individual Missions), regulatory innovations related to specific intervention areas or economic activities, intended to introduce more efficient regulatory and procedural frameworks in their respective sectors;
- competing reforms, i.e., measures not directly included in the Plan but necessary for achieving the overall objectives of the PNRR (such as, for example, tax reform or strengthening the social safety net system).

1.6.2.1 Initiatives and Opportunities Within PNRR

Digital Innovation in Italy: The National Rehabilitation and Resilience Plan defines objectives and interventions that span various sectors of Italy's economic and social spheres. From Tourism to Culture, from Agriculture to Logistics, from Education to Employment, from Healthcare to Public Administration. The common denominator within the Reforms and Missions defined in the Plan is Digital Innovation. The Plan dedicates an entire mission (Mission 1) to digital innovation, one of the six missions it is structured around. Over 40 billion euros, in addition to digitalisation initiatives in other missions, bringing the total value to over 65 billion euros. *PNRR and Digitalisation of the Public Administration*: The Public Administration (PA) plays a central role in the implementation of the PNRR. It is enough to note that 60% of the resources allocated to the Plan are destined for central PA (such as INPS), local PA (such as Municipalities), or public enterprises (e.g., Trenitalia), and all the resources are managed by Ministries or Departments of the Presidency of the Council, i.e., the PA. The PNRR dedicates an entire component of Mission 1 to the Public Administration, specifically aimed at the digital transition of the PA. This area includes various reforms and investments aimed at completing many initiatives already started with the Three-Year Plan and accelerating efforts in historically underdeveloped areas, such as the Cloud migration of public services, digital skills, and the simplification of administrative and judicial processes.

PNRR and Healthcare: Mission 6 of the PNRR dedicates a significant section to Health and Healthcare reforms and investments, particularly aimed at strengthening and developing territorial networks and healthcare services, as well as enhancing digital skills in the doctor-patient relationship and the use of tools such as the Electronic Health Record and Telemedicine. This represents a unique moment in the history of investments and changes in the Italian Healthcare System aimed at innovating the National Health Service.

PNRR and Tourism: Among the numerous economic sectors involved in the PNRR, tourism is one of the main areas of intervention, with investments aimed at improving infrastructure, services, and offerings. Mission 1, focused on the digital transition of the Italian production system, aims to boost tourism and culture through the allocation of 2.4 billion euros. Digitalisation, competitiveness, and the *green* transition are the main objectives pursued by the PNRR.

PNRR and the Primary Sector: Among the PNRR measures, Mission 2 includes the allocation of 5.2 billion euros for the agricultural sector to support technological innovation. Firstly, an upgrade in agricultural machinery is foreseen, which is key to the Agricultural 4.0 market. Additional measures focus on technological innovation in the sector, a transformation that could make a difference for all actors involved in the agri-food supply chain.

PNRR and the Labor Market: Various interventions are aimed at the selection and training of employees within the Public Administration, the development of digital skills necessary to support the *green* and digital transitions, and the reform of active labour policies and vocational training.

PNRR and Digital Identity: In Mission 1, dedicated to digital transition and competitiveness, the Plan provides an integrated offer of digital services aimed at optimizing processes, reducing Public Administration times, and simplifying life for citizens, businesses, and entities. Regarding digital identity, the plan aims to strengthen the system by building on existing platforms (SPID and CIE), before converging towards a new solution with a better user experience and greater accessibility to digital services.

PNRR and Startups: The *startup* ecosystem is also involved in the PNRR plan, although to a limited extent. On one hand, funds are allocated for *startups* and venture capital active in ecological transition within the framework of Mission 2. On the other hand, the Plan seems to focus primarily on other types of investments and reforms mainly related to the Public Administration.

PNRR and Space Economy: Within Mission 1, 1.49 billion euros have been allocated to the Space Economy, the new frontier phenomenon that combines space and digital technologies to develop technological and business opportunities. The space economy encompasses several missions of the PNRR. For example, space technologies can contribute significantly to the *green* revolution and ecological transition sought in Mission 2. Other applications, such as satellite tracking, can support sustainable mobility and logistics (Mission 3) or even Healthcare and Telemedicine (Mission 6).

PNRR and Supply Chain Finance: Within Missions 1, 2, and 3, there are interesting opportunities for the Supply Chain Finance market, a set of innovative solutions that allow a company to finance its working capital. Specifically, three solutions benefit from the PNRR measures: Equipment Finance, Transfer of Tax Credits, and *Deep Tier Finance.*

PNRR and Smart Buildings: To keep pace with the rest of Europe, the PNRR plans to equip more infrastructures, both public and private, with high levels of connectivity, also leveraging the functionalities of other technological applications, including *IoT* systems. Not only advanced sensors applied to production processes, mobility, or logistics, but entire connected ecosystems within private homes and public buildings, often characterized by outdated structures, high maintenance costs, and significant inefficiencies in terms of energy consumption. Transforming these structures into real *Smart Buildings* such as schools, cinemas, and theatres.

PNRR and Smart Cities: From energy efficiency to mobility, from security to the regeneration of urban spaces, to the digitalisation of local authorities: alongside Smart Buildings, the PNRR also permeates the Smart City and *Smart Village* themes. The application areas of the National Rehabilitation and Resilience Plan that fall within the sphere of influence of smart cities and villages are many, and the potential of the interventions is very high, though still largely to be unravelled. Elements of Smart Cities are present in at least three of the six missions defined by the PNRR, particularly in Mission 5, which discusses Integrated Urban Plans, but not only.

The implementing entities at the ministerial level for the cultural investments foreseen by the PNRR are:

- Investiment 1.1: Digital Strategy and Platforms for Cultural Heritage—Central Institute for the Digitalisation of Cultural Heritage—Digital Library;
- Investiment 1.2: Removal of physical and cognitive barriers in museums, libraries, and archives to enable broader access and participation in culture—General Directorate of Museums;
- Investiment 1.3: Improving energy efficiency in cinemas, theaters, and museums—for cinemas, the General Directorate of Museums; for theaters and cinemas, the General Directorate of Cinema and Audiovisual;

- Investiment 2.1: Attractiveness of Historic Villages—Service V—Contracts and Program Implementation of the General Secretariat;
- Investiment 2.2: Protection and enhancement of rural architecture and landscapes—Service V—Contracts and Program Implementation of the General Secretariat;
- Investiment 2.3: Programs to enhance the identity of places: historic parks and gardens—Service V—Contracts and Program Implementation of the General Secretariat;
- Investiment 2.4: Seismic safety in places of worship, restoration of cultural heritage from the Fund for Religious Buildings (FEC) and restoration sites for works of art (Rehabilitation Art)—for seismic safety, the General Directorate for the Security of Cultural Heritage; for the restoration of FEC heritage, the Special Superintendency of Rome; for Rehabilitation Art, Service V—Contracts and Program Implementation of the General Secretariat;
- Investiment 3.1: Development of the film industry (Cinecittà Project)—General Directorate of Cinema and Audiovisual;
- Investiment 3.2: Capacity building for sector operators—General Directorate for Contemporary Creativity.

Investment 2.1—Attractiveness of Historic Villages

Due to the specific nature of this monograph, this investment related to the contribution provided by the PNRR for Italy's small inner areas will be further explored. As part of the PNRR investments, the Ministry of Culture (MiC) has published a public notice for "Proposals for interventions for the cultural and social regeneration of small historic villages," aimed at promoting projects for the regeneration, enhancement, and management of the cultural heritage rich in history, art, culture, and traditions present in Italy's smaller towns. The objectives focus on the protection of cultural heritage, social and economic revitalization, employment recovery, and combating depopulation. This public notice of December 20, 2021, concerns Mission 1—Digitalisation, Innovation, Competitiveness, and Culture, Component 3—Culture 4.0 (M1C3), Measure 2 "Regeneration of small cultural sites, cultural, religious, and rural heritage," Investment 2.1: "Attractiveness of historic villages," funded by the European Union—Next Generation EU. The components of this notice mainly focus on two intervention lines:

1. Intervention Line A: implementation of 21 projects of particular importance and significance, with a budget of 20 million euros each, selected by regions and autonomous provinces.
2. Intervention Line B: implementation of local cultural and social regeneration interventions in at least 229 historic villages.

The recipients of Intervention Line B are small municipalities, either individual or grouped (up to 3), with a population of up to 5,000 inhabitants, in which a recognizable historic village is present. The PNRR resources, distributed among the regions and provinces, are allocated 40% to the South and 60% to the Centre-North.

1.6.2.2 Investment 2.1—Attractiveness of Villages

Investment 2.1—Attractiveness of Historic Towns—Line A: As part of the Investment, but for municipalities with more than 5,000 inhabitants, and thus more populated, 420 million euros are allocated to Line A, which will support pilot projects for the cultural, social, and economic regeneration of villages at risk of abandonment or already abandoned, through the implementation of a limited number of exemplary interventions, one for each Region or Autonomous Province, totalling 21 projects and a budget of 420 million euros. Each intervention will amount to 20 million euros and is aimed at the economic and social revitalization of depopulated villages or those characterized by an advanced process of decline and abandonment. The projects include the establishment of new functions, infrastructures, and services in the fields of culture, tourism, social services, or research, such as schools or academies of arts and cultural crafts, widespread hotels, artists' residencies, research centres and university campuses, residential care homes (RSA) where cultural programs will also be developed, residences for families with workers in smart working and digital nomads. For this line of action, Regions and Autonomous Provinces submitted their proposals to the Ministry of Culture (MiC); a negotiation phase followed, conducted by a Technical Committee established by MiC, which included a representative from the Regions, a representative from ANCI, and a representative from the associations participating in the coordination committee for villages. The negotiation process, aimed at verifying the consistency of the project proposals with the processes and timelines of the PNRR, concluded with the admission of 20 proposals for funding: *Abruzzo: Rocca Calascio (AQ); Basilicata: Monticchio Bagni (PZ); Calabria: Gerace (RC); Campania: Sanza (SA); Emilia-Romagna: Campolo (BO); Friuli-Venezia Giulia: Borgo Castello (GO); Lazio: Trevinano (VT); Liguria: Borgo Castello (SV); Lombardia: Livemmo (BS); Marche: Montalto delle Marche (AP); Piemonte: Elva (CN); Puglia: Accadia (FG); Sardegna: Ulassai (NU); Sicilia: A Cunziria (CT); Toscana: Borgo di Castelnuovo in Avane (AR); Umbria: Cesi (TR); Valle d'Aosta: Arvier (AO); Veneto: Recoaro Terme (VT); Provincia di Bolzano: Stelvio (BZ); Provincia di Trento: Palù del Fersina (TN).*

Investment 2.1—Attractiveness of Villages—Line B: The Notice was aimed at small municipalities, either individual or aggregated, with a resident population of up to 5,000 inhabitants, where a clearly identifiable and recognizable historic village was present. Aggregations of municipalities could involve neighbouring municipalities or those within the same region that shared similar themes. In the case of aggregations of more than one municipality, up to a maximum of three, each municipality had to have a historic village. For aggregated municipalities, the application had to be submitted by the municipality acting as the proposer and lead partner. Line B, aimed at supporting Local Projects for Cultural and Social Regeneration, received a total funding of 580 million euros. Additionally, 20 million euros were allocated to the "Roots Tourism" initiative, managed by the Ministry of Foreign Affairs and International Cooperation. It benefited from a financial allocation of 380 million euros and was aimed at the implementation of projects in at least 229 historic villages, in line with the target set in the investment 2.1 PNRR-M1C3-Culture document. Subsequently, through a separate procedure and with a financial allocation of 200 million euros, enterprises operating in the cultural, tourism, commercial, agri-food, and craft sectors in the same municipalities targeted by the cultural and social regeneration projects were centrally supported under the responsibility of the Ministry of Culture. The public notice, dedicated to small historic villages, aimed to promote projects for the regeneration, enhancement, and management of the rich heritage of history, art, culture, and traditions present in small Italian towns. This initiative integrated the goals of protecting cultural heritage with the needs of social and economic revitalization, job creation, and combating depopulation. The eligible projects for funding had to contribute to the renewal of the socio-economic fabric of these areas by regenerating public spaces, restoring historic-architectural heritage, and activating entrepreneurial and commercial initiatives capable of generating employment in the territory.

Among the projects funded by the PNRR, two initiatives stand out for their significant contribution to innovation.

Innovative Schools: The "Innovative Schools" project of the Ministry of Education, aimed at renovating school buildings and equipping them with new technologies, has received an investment of 1.5 billion euros. This investment is being used for renovation, safety improvements, and the installation of new technologies in 2,000 schools across Italy. The main objectives of the project include improving the quality of education using innovative technologies, such as distance learning, virtual reality, and augmented reality; making schools safer and more welcoming through building renovations and safety measures; and promoting inclusion and participation for students with disabilities or learning difficulties. The project has already been completed in about 500 schools and shows the first encouraging results, highlighting how new technologies can improve student learning and make schools more inclusive.

Renewable Energy: The "Renewable Energy" project by the Ministry of Ecological Transition aims to promote the use of clean and sustainable energy. An investment of 19.4 billion euros has been used to develop solar, wind, and geothermal energy production plants. The main objectives of the project include reducing dependence on fossil fuels and cutting *green*house gas emissions; creating new jobs in the renewable energy sector; and contributing to the decarbonization of the Italian economy.

The project has led to the creation of over 1,000 new renewable energy production plants, helping to reduce Italy's dependence on fossil fuels and improve air quality. These are just a few examples of the many projects and initiatives that have been implemented or are underway as part of the PNRR.

1.6.2.3 PNRR Culture: Enhancement of Architecture and Rural Landscape

In the context of the funding outlined by the PNRR, particular attention is given to rural architecture, with the goal of recovering the entire landscape, overcoming the limited focus on the historic centres of small inner towns in Italy, and embracing the landscape, including isolated heritage sites. The importance of rehabilitating rural architecture plays a crucial role in creating a harmonious connection between the historic centre and the surrounding landscape. It is an inseparable synergy of history, environment, and anthropic landscape. The concept of landscape here takes on a broader meaning, characterized as a seamless fusion between nature and the man-made. The proposal promotes a systematic process of identification, conservation, and enhancement of a diverse heritage, understood as "villages." The primary objective is to preserve the uniqueness of the rural landscape while simultaneously contributing to the protection of cultural heritage and stimulating local development.

The challenges faced are numerous:

- preserve the distinctive elements of historic rural landscapes, with particular attention to the specific distribution of scattered buildings; numerous structures have suffered from abandonment and decay, putting at risk their typological features and their connection with the surrounding spaces;
- reduce regional disparities in the country, promoting the creation of employment and mitigating the social and economic impact of the pandemic crisis: the intervention aims to strengthen rural economies and support tourism in inner areas, enhancing products linked to agriculture and traditional craftsmanship.

In addition to generating employment in the agricultural and construction sectors, the proposal acts as a counter-cyclical measure by involving operators and professionals specialized in landscape and architectural rehabilitation, focusing on vernacular architecture and its integration into the environmental fabric. The main objectives of the investment include preserving the values of historic rural landscapes within protected areas, safeguarding and enhancing cultural heritage (both tangible and intangible), and promoting traditional agricultural practices essential for maintaining the uniqueness of the landscape while encouraging initiatives related to sustainable tourist-cultural use, respecting local traditions and culture, and adapting them to local specificities. The total investment amounts to 600 million euros, of which 590 million are allocated for the restoration of agricultural settlements and historic rural buildings, including scattered and isolated ones. The remaining 10 million euros are dedicated to activities for cataloguing rural heritage assets and implementing information systems.

The initiative, coordinated by the Ministry of Culture, has already achieved successes with numerous interventions, although the results may have been more significant. This step is of considerable importance, especially considering that, beyond the picturesque villages, Southern Italy and Sicily host a range of isolated heritage assets within the landscape. This incentive not only aims to restore rural architecture but also serves as a valuable ally in preserving and enhancing a vast heritage. The initiative adds value to the territory by contributing to the preservation of historical and cultural traces that characterize particularly the smaller towns. The diversity and spread of these assets, scattered across the territories, add authenticity and charm to this corner of Sicily, making it even more attractive for residents and visitors.

The investment aims to enhance historic rural buildings with the goal of increasing tourist flow to the peripheral areas of the country, as well as improving the protection of the surrounding landscape. Through the rehabilitation of abandoned or aging buildings, it is expected to create new jobs, reduce regional disparities, and mitigate the impact of the crisis. This process will also help enhance local craftsmanship and promote agricultural products from the region. This investment aims to promote a systematic process of valorising historic rural buildings, both private and those belonging to third-sector entities, while protecting the landscape. Numerous rural buildings and agricultural structures have undergone gradual abandonment, decay, and alterations, compromising their distinctive characteristics and their relationship with the surrounding spaces. By recovering the rural building heritage, the intervention will contribute to improving the landscape quality of the national territory, returning to the community an underutilized building heritage that is currently inaccessible to the public. The positive implications of this intervention will reflect on local economies, promoting sustainable tourism in rural areas and enhancing production linked to agriculture and traditional craftsmanship.

1.6.3 National Plan for the Revitalization of Small Municipalities

The National Plan for the Revitalization of Small Municipalities is a public investment plan aimed at supporting sustainable development, demographic balance, the protection and enhancement of natural, rural, historical-cultural, and architectural heritage, as well as the system of essential services in small municipalities in Italy. The plan was approved by the Council of Ministers on July 20, 2022, and provides a total investment of 1.3 billion euros for the period 2022–2026. This important initiative aims to support the development of Italy's small municipalities, which represent a significant portion of the country's territory and population. The beneficiaries of the plan are small municipalities in Italy, defined by a population of fewer than 5,000 inhabitants. Projects must be submitted by municipalities individually or in association, and eligible projects relate to interventions in: urban regeneration and enhancement of real estate heritage; promotion of economic and social development; protection and enhancement of the environment and improvement of public services.

Among the projects that can be funded by the Plan:

- revitalization of historic centers: projects for the revitalization of historic centers may include interventions for the conservation and restoration of historic buildings, the creation of public and pedestrian spaces, and the promotion of cultural and tourist activities;
- rehabilitation of abandoned public buildings: abandoned public buildings can be recovered to host social, cultural, and economic activities or be converted into housing;

- development of infrastructure for sustainable mobility: infrastructure for sustainable mobility, such as bicycle paths, hiking trails, and local public transport, can contribute to improving quality of life and reducing pollution;
- promotion of tourism and cultural activities: tourism and cultural activities can help create jobs and promote local economic growth;
- investments in agriculture and rural development: investments in agriculture and rural development can help preserve the environment and promote the production of local food.

The plan represents an opportunity for small municipalities to improve their service offerings and attractiveness, aiming to attract new residents and economic activities. Beyond the national calls for proposals promoted, *Invitalia*, an Italian government agency, aims to foster the economic and social development of the country. Among its various initiatives, one of the most significant is its commitment to promoting calls for proposals for the revitalization of Italy's small municipalities. These calls are designed to provide financial incentives and technical support to municipalities seeking to undertake projects for the renewal and improvement of their communities. A key aspect of this commitment is the allocation of funds for projects involving infrastructure, public services, tourism, culture, and much more. *Invitalia* collaborates with local authorities to identify the specific needs of small municipalities and develop targeted action plans. These plans aim to improve the quality of life in these towns, making them more attractive places for residents and visitors. The promoted calls often offer opportunities for funding and technical support for a wide range of projects. These projects may include the redevelopment of urban areas, the promotion of local tourism, the development of cultural initiatives, and the creation of new local businesses. The goal is to stimulate economic growth, create jobs, and preserve Italy's cultural heritage in small municipalities.

1.6.4 SNAMI—National Strategy for Italian Mountain Areas

The National Strategy for the Italian Mountains (SNAMI) is an initiative funded through the Fund for the Development of Italian Mountains (FOSMIT). This fund gathers resources from the National Mountain Fund and the Supplementary Fund for Mountain Municipalities, with an allocation of €100 million for 2022 (compared to the previous €29.5 million) and €200 million starting from 2023, as established by the recent budget law. The significant increase in funds will ensure coverage for the planned support measures. This was outlined in a draft law approved by the Council of Ministers to promote the development of the country's mountainous areas.

Specifically, incentives are provided for doctors and social-healthcare workers employed in mountain municipalities, and privileged access is granted to the role of medical director for doctors who have practiced in mountain areas. Tax benefits are also available for operators who rent or purchase properties in mountain municipalities. The Council of Ministers, on the proposal of then Prime Minister Mario

Draghi and the Minister for Regional Affairs and Autonomies, *Mariastella Gelmini*, approved the draft law titled "Provisions for the Development and Valorisation of Mountain Areas", which introduces comprehensive measures aimed at fostering economic development and the revitalization of many areas that have the opportunity to increasingly become a resource for the country. It also aims to counter depopulation in Italian mountain areas, consolidating legislative interventions into a single text to reduce the disadvantaged conditions of mountain municipalities. Finally, it establishes criteria for identifying mountain municipalities and regulates the additional parameters for accessing the incentives and support provided. Through SNAMI, the Minister for Regional Affairs and Autonomies identify the strategic guidelines for economic and social growth, accessibility to essential services and digital infrastructure, and the full enjoyment of fundamental rights in mountain areas. The measures adopted by the draft law specifically address:

- mountain healthcare: incentives are provided for doctors and social-healthcare workers employed in mountain municipalities. Specifically, privileged access is granted to the position of medical director for doctors who have practiced in mountain areas. Additionally, a tax credit is granted to doctors and social-healthcare workers who rent or purchase a residential property in mountain municipalities eligible for such benefits;
- mountain schools: facilitating measures and incentives are planned for teachers working in mountain areas. Incentive schemes, including service score benefits, are introduced for temporary teachers in mountain schools at all levels. Furthermore, mountain teachers are granted a tax credit for housing under the same conditions provided for doctors and social-healthcare workers in mountain areas;
- mobile phone services and internet access: High-speed internet access and infrastructure improvements are encouraged to ensure continuity of mobile phone services. Measures under the National Rehabilitation and Resilience Plan (PNRR) are further strengthened, extending their reach to territories excluded from the PNRR;
- incentives for agricultural and forestry entrepreneurs: a contribution in the form of a tax credit is provided to agricultural and forestry entrepreneurs operating in mountain municipalities who invest in environmentally and climate-beneficial practices. The specific practices for agricultural entrepreneurs will be identified by a decree from the Minister of Agricultural, Food, and Forestry Policies. For forestry entrepreneurs, these practices include crop diversification, maintaining or introducing areas of ecological interest, and preserving permanent grasslands;
- fiscal measures for "young" mountain enterprises: a tax credit is introduced for the first three fiscal periods, targeting small and microenterprises starting their business in mountain municipalities eligible for this incentive, where the owner or at least one of the proprietors is under 36 years of age.

"*Io Resto in Montagna*" measure: this measure aims to encourage residency in mountain areas. Taxpayers under the age of 41 are granted a higher tax deduction than the standard one for interest payments on mortgages taken out for the purchase of their primary residence. The deduction applies to mountain municipalities with a resident population not exceeding 2,000 inhabitants that are eligible for this form of benefit. The program was presented as part of the bill "Provisions for the Development and Enhancement of Mountain Areas", approved by the Council of Ministers on July 20, 2022. This initiative aims to encourage residency in mountainous regions by offering tax benefits that exceed those normally provided for interest payments on mortgages taken out for the purchase of primary homes in mountain municipalities with a resident population not exceeding 2,000 inhabitants. The tax deduction offered amounts to 30% of the interest paid over the first three fiscal periods, with a maximum limit of 1,500 euros per fiscal period. This tax benefit can be combined with other incentives available for purchasing a first home.

This initiative holds significant relevance in addressing the depopulation phenomenon affecting mountain regions. The tax incentive serves as a tangible encouragement for young people and families to choose to settle in these areas, where they can enjoy a unique natural environment and a quieter atmosphere. Although the project is still in the implementation phase, it has already sparked considerable interest. According to a survey conducted by the Ministry for Regional Affairs and Autonomies, 40% of young Italians express interest in living in mountainous areas. "Io Resto in Montagna" represents just one of the measures included in the National Strategy for the Italian Mountains (SNAMI) to counter depopulation. Other measures include incentives for doctors and social-healthcare workers operating in mountain areas, tax benefits for agricultural and forestry entrepreneurs, and funding aimed at developing infrastructure. The implementation of SNAMI represents a significant opportunity to revitalize mountain areas and make them more attractive to young people and families, contributing to the overall well-being of Italy's mountainous regions.

1.6.5 SNAC—National Strategy for Climate Change Adaptation

On October 30, 2014, the Ministry of the Environment launched the National Strategy for Climate Change Adaptation [45], a significant document providing a national strategic vision on addressing the impacts of climate change as defined in the 2013 European Climate Change Adaptation Strategy, subsequently revised in 2021. The Strategy was drafted by the Ministry of the Environment and the Protection of Land and Sea, with scientific coordination from the CMCC, and received approval from the Unified Conference. The Strategy's implementation tool is the National Plan for Climate Change Adaptation, which defines actions and priorities at both national and local levels to mitigate the impact of climate change on the environment, socio-economic sectors, and natural systems in Italy. It identifies the primary impacts of climate change on environmental resources and significant national socio-economic sectors, proposing initial adaptation actions for each. Its aim is to minimize risks posed by climate change, protect public health, well-being, and assets, preserve natural heritage, maintain or improve the resilience and adaptive capacity of natural, social, and economic systems, and leverage potential opportunities arising from new climatic conditions. To achieve its objectives, the Strategy outlines the following actions: improving current knowledge of climate change and its impacts; assessing territorial vulnerability and adaptation options for natural systems and relevant socio-economic sectors, as well as associated opportunities; promoting participation and raising awareness among *stakeholders* through extensive communication and dialogue to integrate adaptation effectively into sectoral policies; supporting awareness and information campaigns on adaptation through widespread communication about possible hazards, risks, and opportunities arising from climate change; and specifying tools to identify the best adaptation options while emphasizing co-benefits. In addition to serving as a reference framework for adaptation for regions and municipalities, the SNAC provides objectives, principles, and a set of sectoral and cross-sectoral adaptation actions (in areas such as water resources, agriculture, health, transport, energy, etc.). This marks progress towards sustainable development in Italy, offering new tools to prevent and manage climate risks in key sectors and to generate added value from climate change knowledge. The Strategy outlines four specific objectives: reducing the vulnerability of natural, social, and economic systems to climate change impacts; increasing their adaptive capacity; enhancing the exploitation of opportunities; and coordinating actions at various levels. Supporting documents include: "Report on the state of scientific knowledge on climate impacts, vulnerability, and adaptation in Italy", "Analysis of relevant European and national regulations on impacts, vulnerability, and climate adaptation", and "Elements for a National Strategy for Climate Change Adaptation". Annexes to the SNAC include: an in-depth chapter on the European Strategy for Climate Change Adaptation and strategies and plans adopted in other European countries; a set of proposed actions categorized as non-structural actions, ecosystem-based actions, infrastructural and technical actions, and cross-sectoral actions, both short- and long-term. The SNAC represents a key response by Italy to the global challenges posed by climate change. This strategy aims to protect communities, natural resources, and the national economy through climate adaptation measures, a critical component particularly for Italy's small, inner rural areas, which serve as repositories of the country's heritage. It is important to emphasize the need for targeted action, particularly in Southern Italy, where rising temperatures have had a significant impact on economic performance.

1.6.6 PAC 2023–2027—Action and Cohesion Plan

In Europe, one of the main topics under discussion regarding the new PAC 2023–2027 Programming is related to the development of so-called *Smart Villages*, seen as a strategic tool to strengthen the socio-economic fabric of rural areas. The topic of *Smart Villages* is not a completely new concept introduced by the upcoming five-year community programming period. As early as 2017, the European Union developed the strategic document *EU Action for Smart Villages*, within which villages, or more commonly hamlets, are defined as "*communities located in rural areas that use innovative solutions to improve their sustainability and resilience, based on local strengths and opportunities*" Therefore, it is not a universal model to be applied indiscriminately everywhere but rather an approach that must be contextualized to each territory and should be integrated into broader regional and/or territorial development strategies (such as, for example, the LEADER Local Development Strategies).

The European Commission has given the PAC 2023–2027 a strategic role in promoting *Smart Villages*, providing support for "*Smart Village*" strategies among the cooperation interventions outlined in Article 77 of Regulation (EU) 2021/2115. To further emphasize the importance of this programming approach, the Commission has established a specific result indicator for the PAC Strategic Plans: the indicator R.40 Smart Transition of the Rural Economy, which aims to assess the number of "Smart Village" strategies funded through the PSPs. This type of intervention primarily contributes to specific objective 8 "*Promote employment, growth, gender equality, including women's participation in agriculture, social inclusion, and local development in rural areas, including circular bioeconomy and sustainable forestry*" and represents an additional tool to encourage aggregated territorial and local approaches.

Other Member States have taken different approaches, with *Smart Villages* being the exclusive prerogative of LEADER or one of the investment forms provided for under Articles 73 and 74 of Regulation EU 2021/1060. However, not all Member States have explicitly referred to *Smart Village* strategies in their PSPs, and even fewer have attempted to provide guidance on the targets to be achieved regarding their implementation; among these are Spain, Germany, Austria, Finland, and Italy.

Spain is one of the countries that has chosen to include *Smart Village* strategies among the possible projects financed through the intervention "Non-productive investments in basic services in rural areas." As in Italy, a Member State with regionalized PAC implementation, this intervention will be carried out by 11 of the 17 autonomous communities into which the country is divided. Galicia, for example, intends to finance investments for the provision of telecommunications services, from purchasing equipment to developing the necessary software for proper service delivery, as well as planning, management, or any other activity that contributes to the objectives set in the digitalisation plans. The Valencian Community, on the other hand, has chosen to focus its concept of *Smart Village* on projects aimed at energy independence with renewable energy in municipalities with fewer than 5,000 inhabitants.

Germany has also chosen to direct its support for the smart transition of the rural economy towards the field of digitalisation, to provide rural businesses and populations with adequate access to modern information and communication technologies: the German Managing Authority has planned the implementation of eleven operations for broadband expansion.

Austria, however, has adopted a different approach to the *Smart Village* concept, preferring to identify several interventions that contribute to the implementation of *Smart Village* strategies through a bottom-up approach and integrated planning, thanks to cooperation between local actors and not extended to the Local Action Groups (LAGs). With interventions 73–10—aimed at promoting urban centres, and 77-4 for the revitalization of vacant properties, Austria intends to fund investments for the regeneration and rehabilitation of public areas and/or buildings that are abandoned, underused, or misused. Additionally, there is an effort to raise awareness among citizens, community representatives, and economic operators on strengthening urban centres by financing the preparation and updating of integrated urban development plans for the reuse of revitalized spaces. Another cooperation intervention called "Rural Innovation Systems" aims to fund cooperation strategies for innovation in non-agricultural areas and an innovation network that synergizes partnerships and their respective projects. Finner has established an intervention called "*Smart Village* Cooperation Projects" identifying three main themes within which beneficiaries can develop their strategies: economic innovation, the social sphere, and environmental protection. For each of the three action areas, a wide range of sectors is identified in which each strategy can focus: transport, smart farming, training, food districts, etc. Other Member States, such as Poland, Slovenia, and Estonia, plan to support *Smart Village* strategies in various ways, particularly focusing on innovation and digitalisation, and above all, aligning with the LEADER approach.

Italy has chosen to build the intervention by interpreting it, as suggested by the CE Tools, as a methodological approach to support the implementation of "strategies" understood as integrated projects, shared by public and/or private local beneficiaries, related to specific sectors/areas of cooperation, aiming to promote the use of innovative solutions in rural communities to improve their resilience, economic, social and/or environmental conditions, local *governance*, relationships, and exchanges with urban areas, also utilizing digital technologies and multifunctional agriculture and forestry. Moreover, in Italy, *Smart Villages* can be incentivized both through LEADER Local Development Strategies and through regional calls for projects.

1.7 Codes and Directives for the Protection and Enhancement of Historic Centres

The protection and enhancement of historic centres, particularly in the picturesque smaller towns of Italy, represents a crucial challenge in the context of cultural heritage conservation. Alongside the numerous initiatives and regulations issued at both the Italian and European levels, as outlined in the previous paragraphs, the promotion of local heritage also includes restoration charters, Italian regulations, guidelines for the protection and enhancement of inhabited centres, and several European directives and conventions. The Restoration Charters are essential documents that outline principles and guidelines for interventions and play a fundamental role in this process, emphasizing the importance of preserving the authenticity and specificity of these places, such as the small inner towns, often composed of dense historic centres rich in history, valuable architecture, culture, and tradition. The Restoration Charters, such as the first and most famous, the Athens Charter of 1931, and the most recent, the Krakow Charter of 2000, provide a regulatory and ethical framework for the restoration of historic centres. These documents highlight the importance of a careful approach, respectful of the original characteristics, architectural peculiarities, and historical layers present in the villages, avoiding interventions that could compromise their historical and architectural integrity. In Italy's most authentic smaller towns, the application of the Restoration Charters translates into a thorough analysis of local architectural and cultural characteristics before undertaking any intervention. The goal is to preserve the traditional urban fabric, squares, cobblestone streets, and historic buildings, maintaining the original appearance and environmental context, thereby promoting local identity. The objective is to reconcile conservation with contemporary needs, allowing historic centres to continue living, adapting, and approaching innovation without losing their authenticity.

In many Italian municipalities, the active involvement of residents in the management and preservation of the historic centre is a key element in ensuring the success of rehabilitation initiatives, based on the awareness of each place's uniqueness and the desire to preserve collective memory. Through a thoughtful and respectful approach, these places become not only testimonies of the past but also foundations for the future. Italian legislation for the conservation of cultural heritage has a long history, rooted in the nineteenth century. In 1882, with Law No. 364, the Superintendence for Monuments and Fine Arts was established to protect and enhance Italy's cultural heritage.

In 1939, with Law No. 1089, the concept of "cultural heritage" was introduced, encompassing immovable and movable properties of historical, artistic, archaeological, ethnographic, archival, and bibliographical interest. The law also established that cultural heritage was subject to protection by the State, which had the right to intervene when necessary. In 1974, with Law No. 1089, the Superintendence for Environmental and Architectural Heritage was established, tasked with protecting, supervising, and enhancing environmental heritage, that is, immovable properties of landscape, historical, artistic, architectural, archaeological, ethnographic, and scientific interest. In 1999, with Law No. 368, the concept of "enhancement of cultural heritage" was introduced, which included not only protection but also the promotion and enjoyment of cultural heritage. The law also established the creation of a fund for the enhancement of cultural heritage. In 2004, with Law No. 42, the Code of Cultural Heritage and Landscape was approved, which simplified and rationalized the regulations regarding the protection and enhancement of cultural heritage. The Code also introduced new provisions, such as the possibility of assigning restoration work to private entities through concessions or contracts. Italian legislation on conservation and restoration has evolved over time, responding to new needs for the protection and enhancement of cultural heritage. In recent years, there has been a growing interest in the enhancement of cultural heritage, which is increasingly seen as an essential element for its protection. Historic centres represent an invaluable cultural heritage that testifies to the history and culture of our country. These centres are often characterized by valuable, authentic architecture, making these places unique in the world.

However, the historic centres of small villages are also subject to numerous risks, including degradation, abandonment, and mass tourism. To safeguard these centres, coordinated action by institutions is required, based on a solid regulatory foundation and perfect synergy between the fields of rehabilitation and restoration.

The restoration charters and Italian laws play a fundamental role in safeguarding and provide a framework to ensure that these interventions are carried out in compliance with scientific principles and the protection of cultural heritage. These regulations allow for the integration of cultural heritage protection with the enhancement of these places, which represent an important resource for our country. The study and analysis of the regulatory framework provides a range of interventions and possible actions in well-preserved and protected contexts, such as most of the smaller towns. While focusing exclusively on rehabilitation, such studies are preparatory for understanding territorial dynamics as well as possible actions. In fact, small towns are often located within natural sites recognized worldwide and, in addition to their basic architectural fabric, also possess monumental heritage that makes them unique and important for Italian history.

1.7.1 Davos Declaration 2018. A Culture of Quality Building for Europe

In January 2018, European culture ministers adopted the Davos Declaration titled "Towards a Culture of Quality Construction for Europe." This declaration outlined a strategic and political path to spread and consolidate a culture of high-quality construction. It emphasizes the concept that building is a cultural act that creates space for culture itself. The Davos Declaration has been cited and widely discussed worldwide, becoming a point of reference for states, organizations, institutions, and businesses [46]. The movement for the culture of construction, following the commitment established by the Davos Declaration, continues a strategic, political, and scientific level. It aims to deepen scientific reflection on fundamental concepts and clarify the key terms of the declaration. At the same time, it promotes political strategies that recognize the cultural value of conservation, land planning, and construction, with the primary goal of creating a culture of quality construction [47–49]. Various international initiatives associated with this process are consolidated under the name of the "Davos Process." In small towns that preserve unique heritage and construction traditions, it is imperative to talk about quality. The culture of construction is defined as an approach that considers the built environment, embracing aspects ranging from building substance to contemporary production, from artisanal details to buildings and open spaces, to large infrastructures. The definition of "quality construction culture" remains a subject of debate, but numerous initiatives have been promoted to improve the quality of the built environment.

The Davos System for Quality in the Culture of Construction was developed to assess the qualities of places through eight fundamental criteria: *governance*, functionality, environment, economy, diversity, context, genius loci, and beauty. This tool is useful in planning, projects, competition evaluations, or participatory processes. It is designed to consider and weigh social, emotional, and cultural values alongside technical and functional aspects. The Davos System evaluation module includes a series of questions for each of the eight criteria, adaptable to the specific characteristics of places or projects. The responses allow the determination of the actual quality level of a place, highlighting strengths and weaknesses. The Swiss Foundation for the Culture of Construction responded cautiously to the first version of the Davos System, criticizing it as a "technocratic approach" that could "corrupt the essence of Baukultur." However, the Davos System aims to overcome divisions between disciplines and engage citizens in the dialogue on quality in construction. It also seeks to balance the authority of professionals in judgments through its use in competitions and bidding processes. The challenge is to allow the emergence of a "discourse on quality" that involves not only a predetermined jury but also citizens and professionals outside narrow circles. The goal is to stimulate reflection on the quality of the built environment, seeking to involve designers, citizens, and professionals in the dialogue. The overall evaluation of the Davos System could produce heterogeneous results, influenced by the perspective of the evaluator, which could be the mayor of a small town, as in the case of Italian villages or towns, or an architect. The aim is to initiate independent reflection, allowing each person to ask questions about quality, evaluating their own context and position. Benchmarking and the definition of indicators for each criterion are expected in a later phase.

1.7.1.1 Definition of the Culture of Building Construction

The anthropic environment is part of our daily life, so it is essential to promote a culture of high-quality construction for our overall well-being. Construction culture embraces the anthropic environment in its entirety, considering it as an integrated whole: from the building substance to contemporary production; from small artisanal details to buildings and open spaces, extending to include large infrastructures; from the design phase to the construction phase, from commissioning to reuse, and finally, also considering the landscape. In fact, the precise definition of what falls under the category of "quality construction culture" remains a topic of debate [50]. Numerous initiatives have been undertaken with the aim of improving the overall quality of the anthropic environment. These initiatives range from detailed artisanal activities to the construction of buildings, the creation of open spaces, and the implementation of large infrastructures. In this way, the promotion of a high-quality construction culture aims to optimize every phase of the process, from ideation to reuse, thus contributing to the creation of a more sustainable and rewarding anthropic environment. A culture of high-quality construction is reflected in cities, villages, and buildings conceived and built according to excellence criteria. These quality places contribute to creating a pleasant living environment, promoting the well-being and health of individuals.

The Davos System for quality in construction culture was developed to provide the possibility to evaluate the qualities of places through eight fundamental criteria. Its utility also extends to the planning and design phase, the evaluation of competitions, or in participatory processes. What makes this system unique is its ability to integrate social, emotional, and cultural values with technical and functional aspects, just as it should be for territorial development. Developed in collaboration with international partners by the Swiss Federal Office of Culture, the Davos System for quality in construction culture actively contributes to the Davos Process, initiated in 2018 with the adoption of the Davos Declaration, integrating existing tools and committing to preserve and develop diverse places distinguished by a culture of high-quality construction, also considering three fundamental criteria such as *Firmitas, Utilitas, Venustas*. However, the modern approach stimulates reflection on various projects, multiplying perspectives and encouraging a broader vision. Moreover, the Davos System proposes measuring projects by assigning a score for each category to evaluate them overall. The final evaluation of a project is determined by the average of these scores. According to sociologist Steffen Mau, quantification could lead to a simplification of the criteria and polarization in favor of those that can be expressed numerically.

1.7.2 ICOMOS—Historic Cities and Villages

ICOMOS, the International Council on Monuments and Sites, is an international non-governmental organization and advisory body of UNESCO, with the primary mission of promoting the conservation, protection, use, and enhancement of monuments, built complexes, and sites of historical and cultural significance. The issue of smaller towns plays a crucial role in the development of Italian society, especially considering the acceleration and revival that occurred following the pandemic, which prompted a trend of abandonment of large metropolitan areas in favour of small municipalities, aimed at ensuring a better quality of life. ICOMOS CIVVIH has actively contributed to the integrated urban conservation of historic cities and villages internationally over the years. A significant step in this direction is the drafting of the "Valletta Principles for the Safeguarding and Management of Historic Cities and Urban Areas" [47–49]. In 2020 and 2021, ICOMOS focused its attention on enhancing the rich cultural, material, and immaterial heritage of villages and small towns, emphasizing the need for their regeneration and valorisation.

In this context, the theme of Law No. 158 of 2017, commonly known as the "Save the Villages Law" (further discussed in the next section), was explored, aiming to encourage an improved proposal for its application and ensure its full enforcement. At such a crucial time, when it is necessary to synthesize the various strategic projects launched in the past, ICOMOS intends to offer its expertise. This includes analysing *best practices* carried out locally, as well as documenting the challenges encountered [51]. ICOMOS CIVVIH intends to establish an Observatory on the most interesting activities developed by local communities, with the goal of identifying concrete proposals that can lead to improvements in the current situation. The goal is to contribute to Italy's effective use of Next Generation EU resources to implement regeneration projects for small towns, safeguarding and enhancing the natural, rural, historical-cultural heritage and ancient local traditions. To combat depopulation, the adoption of new operational tools is hoped for, such as a system of essential services (education, healthcare, and transportation) and technological innovation (broadband), within a framework of sustainable development. This requires directing funds towards a proper intervention line for regeneration and valorisation in public interest operations [52]. For this to happen effectively, it is imperative to define and share future action priorities, encouraging the construction of a new, expanded territorial vision. This implies the involvement of local administrations and public authorities, the main beneficiaries, in defining common ideas to guide spending according to agreed guidelines. Moreover, it is crucial to prevent funds from being directed towards privatized operations of personal interest, instead promoting interventions that contribute to the common good and strengthen the capabilities and skills of the staff and organizations involved. The idea of *Smart Villages* is, therefore, at the core of ICOMOS.

1.7.3 "Salva Borghi" Law

Returning to Italy, the "Salva Borghi" Law No. 158/2017—Small Municipalities was conceived to promote and facilitate the sustainable development—economic, social, environmental, and cultural—of small municipalities. This legislation aims to maintain demographic balance in the country by encouraging residence in these communities and protecting and enhancing their natural, rural, historical-cultural, and architectural heritage [53]. The law promotes specific measures to support residents in small municipalities and the local productive activities, with particular attention to the system of essential services. The main goal is to combat depopulation in these areas and encourage tourist influx [54]. Key features include:

- <u>beneficiaries</u>: Municipalities with a population of up to 5,000 inhabitants or municipalities formed following a merger, each with a population of up to 5,000 inhabitants;
- <u>type of funding</u>: 100% grant (non-repayable);
- <u>financial allocation</u>: 160 million euros until 2023;
- <u>opening of the call</u>: The call is expected to open by the end of the year, once the final implementation measures for granting the contribution are defined.

Settling in small municipalities is considered a crucial resource for preserving the territory, especially for activities aimed at combating hydrogeological instability and for the widespread maintenance of common goods [55]. The law identifies small municipalities as those with a resident population of up to 5,000 inhabitants, including municipalities formed following the merger of municipalities with a population of up to 5,000 inhabitants. Priorities for interventions in small municipalities include:

- qualification and maintenance of the territory, with rehabilitation and redevelopment of existing buildings and abandoned areas, as well as actions aimed at reducing hydrogeological risk;
- safety and redevelopment of road infrastructure and public buildings, with particular attention to schools, early childhood centres, and social-care facilities;
- improving energy efficiency of public building stock and promoting renewable energy production and distribution systems;
- acquisition and redevelopment of abandoned land and buildings to support youth entrepreneurship and initiate new tourist and commercial activities;
- acquisition of former tollhouses and disused railway land, with the aim of developing cycle paths and promoting tourism enhancement;
- urban redevelopment of historic centres, with particular attention to ancient villages and abandoned or depopulated historic centres;
- rehabilitation of cultural, historical, artistic, and library assets;
- rehabilitation of mountain pastures to support the production of high-quality meats and cheeses.

The law includes specific articles, such as the establishment of a Fund for the structural, economic, and social development of small municipalities, funding for the protection of the environment and cultural assets, measures to counter the abandonment of properties, promotion of distributed hotels, development of ultra-fast broadband networks, and promotion of local products through short supply chains and zero-kilometre production: a fundamental tool for preserving and enhancing the heritage of rural communities, supporting their growth and attractiveness [56].

1.7.4 FOSMIT—Fund for the Development of Italian Mountain Areas

On September 18, 2024, a call for proposals of 19 million euros was announced for the mountain and partially mountain municipalities (D.A. no. 367/Serv. 6 of 18.09.2024), funded by the Fund for the Development of the Italian Mountains (Fosmit), with resources allocated by the Presidency of the Council of Ministers to rural areas. The call aims to support initiatives for the enhancement and promotion of mountainous areas, strengthening local social networks, reducing depopulation, and improving the attractiveness of these fragile territories. The eligible areas for funding include:

- informatization: projects for digital literacy and inclusion, aimed at stimulating the use of the web and promoting digital participation of the population;
- instrumental assets: acquisition of tools to improve essential services in municipalities;

- heritage enhancement: promotion and redevelopment of existing heritage, both material and immaterial, to increase tourism and cultural attractiveness;
- energy efficiency: targeted interventions to reduce dependence on fossil energy sources in the medium and long term and to improve the energy efficiency of public infrastructures;
- road maintenance: ordinary and extraordinary maintenance interventions of municipal and inter-municipal roads, essential to improve inner mobility and connectivity between different mountain areas.

The current regional councillor for Local Autonomies and Public Administration, Andrea Messina, emphasized that this call for proposals is the result of intense coordination work between various regional departments and the Basin Authority to recover national funds that would have been lost. Messina reiterated that the 19 million euros recovered represent a concrete commitment from the regional government to support and develop mountain municipalities, giving local administrations the freedom to present projects in areas of significant importance for promoting their territories. The main objective of the call is to strengthen the attractiveness of these vulnerable areas and reduce the risk of depopulation. This will be achieved through the enhancement of social networks and the promotion of cultural, environmental, and infrastructural heritage. This approach aligns with the concept of "Roots Tourism," which promotes a tourism offer linked to the memory of places, particularly focusing on local values and historical traditions. Each project must be approved by the local council before submission, under penalty of exclusion from the call. The general objectives of the call are: strengthen the infrastructure and essential services of mountainous areas, as well as improve roadways through ordinary and extraordinary maintenance of municipal and inter-municipal roads; promote digital transition and the use of technologies for social inclusion; enhance natural and cultural heritage; reduce dependence on fossil energy sources and promote more sustainable use of resources and combat depopulation by improving the quality of life in mountainous areas. The call represents a response to the challenges of depopulation and marginalization faced by the Sicilian mountain areas, offering an opportunity to stimulate their economic, social, and tourism development.

References

1. Bonomi, A. (1997). *Il capitalismo molecolare: La società al tempo del lavoro diffuso*. Einaudi.
2. OECD. (2016). *Rural development in OECD countries: Policies and practices*. OECD Publishing.
3. Bonomi, A., & Masiero, R. (2014). *Dalla smart city alla smart land*. Marsilio Editori.
4. OECD. (2018). *The impact of smart villages on rural development*. OECD Publishing.
5. UN-Habitat. (2019). *The state of the world's cities 2018/19: Urbanization and rural transformation*. UN-Habitat.
6. World Economic Forum. (2018). *Smart villages: A new frontier for rural development*. World Economic Forum.

7. Gonzalez-Alvarez, M. C., Chumachenko, Y., & Vanclay, F. (2017). Smart villages: A new paradigm for rural development. *Sustainability, 9*(12), 2245. https://doi.org/10.3390/su9122245
8. Costa, D., Quinteiro, P., & Dias, A. C. (2019). A systematic review of life cycle sustainability assessment: Current state, methodological challenges, and implementation issues. *Science of the Total Environment, 686*, 774–787. https://doi.org/10.1016/j.scitotenv.2019.05.435
9. Antonelli, C. (2019). Smart villages: A new tool for rural development. In *Proceedings of the 13th international conference on information systems for crisis response and management (ISCRAM 2019)* (pp. xx–yy). Springer.
10. World Bank. (2019). *The future of smart villages*. World Bank.
11. United Nations. (2020). *The role of technology in smart villages*. United Nations.
12. Commission, E. (2017). *Investing in rural areas: A European perspective*. Publications Office of the European Union.
13. McKinsey Global Institute. (2019). *Smart villages: A guide to the future of rural life*. McKinsey Global Institute.
14. Commission, E. (2015). *Smart villages: A guide for local governments*. Publications Office of the European Union.
15. Chiodo, E., Adriani, H. L., Navarro, F. P., & Salvatore, R. (2019). Collaborative processes and collective impact in tourist rural villages—Insights from a comparative analysis between Argentinian and Italian cases. *Sustainability, 11*(2), 432. https://doi.org/10.3390/su11020432
16. ISTAT. (2019). *Le aree interne in Italia*. ISTAT.
17. Union, E. (2018). *Smart villages: A new approach for rural development*. European Commission.
18. Union, E. (2021). *The challenges of smart villages*. European Union.
19. World Bank. (2016). *Smart villages: A handbook for practitioners*. World Bank.
20. Commission, E. (2016). *European broadband strategy 2016–2025*. Publications Office of the European Union.
21. United Nations. (2017). *Smart villages: A case study of success*. United Nations.
22. Al Nuaimi, S. A., Al Neyadi, H., Mohamed, N., & Al-Jaroodi, J. (2018). Applications of big data to smart cities. *Journal of Internet Services and Applications, 9*(1), 11. https://doi.org/10.1186/s13174-018-0087-5
23. Recommendation of the European Commission of 15 February 2016 on the definition of "smart city" and the promotion of innovative approaches to sustainable urban development.
24. Commission, E. (2010). *Digital agenda for Europe*. Publications Office of the European Union.
25. Rossi, F., Rossi, S., & Viganò, M. (2017). *The future of smart villages in Europe*.
26. Dassori, E., Morini, A., Messico, A., Vite, C., Polverino, S., & Morbiducci, R. (2021). A smart village model for the Italian coastal territory. *Rivista Tema, 05*. https://doi.org/10.30682/tema0502j
27. Ratti, C., Cucinelli, M., & D'Orazio, G. (2016). *Smart villages in Europe: A policy perspective*.
28. Caragliu, A., Del Bo, C., & Nijkamp, P. (2009). *Smart cities in Europe* (p. 18). VU University Amsterdam, Faculty of Economics, Business Administration and Econometrics, Serie Research Memoranda. https://doi.org/10.1080/10630732.2011.601117; Baldini, U. (1988). *Manuale del restauro*. Nardini Editore. ISBN: 9788840430226.
29. Panciszko-Szweda, B. (2021). The reasons of the implementation of the concept of smart villages in the European Union. *Przegląd Politologiczny*, 37–48. https://doi.org/10.14746/pp.2021.26.4.2
30. ISTAT. (2023). *Rapporto annuale 2023: La situazione del Paese*. ISTAT. https://www.istat.it/it/archivio/284858
31. Bacilieri, C. (2025). *I borghi più belli d'Italia: Il fascino dell'Italia rivelata. Guida 2025*. Società Editrice Romana.
32. Ciuffetti, A. (2019). *Appennino. Economie, culture e spazi sociali dal medioevo all'età contemporanea*. Carocci.
33. Dematteis, G. (2013). Montagna e aree interne nelle politiche di coesione territoriale italiane ed europee. *Territorio, 66*, 7–15.

34. Dematteis, G. (a cura). (2011). *Montanari per scelta. Indizi di rinascita nella montagna piemontese*. Franco Angeli.
35. Dematteis, G. (2009). Polycentric urban regions in the Alpine space. *Urban Research and Practice, 1*, 18–35; Becattini, G. (2000). *Il distretto industriale: Un nuovo modo di interpretare il cambiamento economico*. Rosenberg & Sellier.
36. Directorate-General for Communication. (2022). *Annual activity report 2021—Communication*. European Commission. https://commission.europa.eu/publications/annual-activity-report-2021-communication_en
37. European Network for Rural Development. (2023, March 25). *ENRD homepage*. European Commission. https://enrd.ec.europa.eu/
38. Convention for European Economic Cooperation. (1948, April 16). *Road to recovery, The Marshall plan: Its importance for the Netherlands and European cooperation*. Ministry of Foreign Affairs, Directorate-General for the Economic and Military Aid Program. http://www.cvce.eu/obj/convention_for_european_economic_cooperation_paris_16_april_1948-en-769de8b7-fe5a-452c-b418-09b068bd748d.html
39. United Nations. (1966). *International covenant on economic, social and cultural rights* (GA Res. 2200A [XXI], 16 Dec. 1966). United Nations. https://www.ohchr.org/en/instruments-mechanisms/international-covenant-economic-social-and-cultural-rights
40. European Commission, Directorate-General for Financial Stability, Financial Services and Capital Markets Union. (2018, March 8). *Renewed sustainable finance strategy and implementation of the action plan on financing sustainable growth*. https://finance.ec.europa.eu/publications/renewed-sustainable-finance-strategy-and-implementation-action-plan-financing-sustainable-growth_en
41. Leuba, P. (2017). *A better future for Europe's rural areas*. Congress of Local and Regional Authorities, Council of Europe.
42. Nieto, E., & Brosei, P. (2019). The role of LEADER in smart villages: An opportunity to reconnect with rural communities. In A. Visvizi, M. D. Lytras, & G. Mudri (Eds.), *Smart villages in the EU and beyond*. Emerald Publishing.
43. Wolski, O. (2019). Digitalisation of rural areas and agriculture in the EU debate: How far from what research says? *Wieś i Rolnictwo*.
44. Visvizi, A., et al. (2019). *Smart villages in the EU and beyond*. Emerald Publishing.
45. SNAC, Strategia Nazionale di Adattamento ai cambiamenti climatici. (2015). www.minambiente.it/notizie/strategia-nazionale-di-adattamento-ai-cambiamenti-climatici-0
46. Jokilehto, J. (2017). *A history of architectural conservation*. Routledge.
47. ICOMOS. (2008). *Nara document on authenticity*. https://whc.unesco.org/document/116018
48. ICOMOS. (2013). *Principles for the conservation of heritage sites*. https://historicengland.org.uk/advice/constructive-conservation/conservation-principles/
49. ICOMOS. (2017). *ICOMOS Charter for the conservation and restoration of monuments and sites (1964)*. https://www.icomos.org/en/participer/179-articles-en-francais/ressources/charters-and-standards/157-thevenice-charter
50. Förster, K. (2023). *Undisciplined knowing: Writing architectural history through the environment*.
51. Almagro Vidal, A., Tandon, A., & Eppich, R. (2015). First aid to cultural heritage. Training initiatives on rapid documentation. *International Archives of the Photogrammetry, Remote Sensing and Spatial Information Sciences, XL-5/W7*, 13–19. https://doi.org/10.5194/isprsarchives-XL-5-W7-13-2015
52. Choay, F. (2001). *The invention of the historic monument*. Cambridge University Press.
53. Romano, F. S. (2022). *La legge salva borghi: Un nuovo modello di sviluppo per le aree interne del Paese*. Gangemi Editore. ISBN: 9788849237740.
54. Zuanna, G., Di-Fonzo, T., Girolimetto, D., & Loghi, M. (2024). Exploring the impacts of COVID-19 on births in Italy, 2020–2022. *Population, Space and Place, 30*, e2807. https://doi.org/10.1002/psp.2807

55. Conti, M., & Sivini, S. (2023). Small municipalities attracting rural newcomers and fostering local cohesion: Innovative approaches for rural regeneration in Italy. *Sustainability, 15*(7), 5837. https://doi.org/10.3390/su15075837
56. Caciagli, C., & Porta, D. (2025). *Solidarity as care: Multidimensional inequalities and narratives in Italian cities*. https://doi.org/10.1007/978-3-031-82563-7_2

Chapter 2
Mapping Smart Village Initiatives: A Comparative Case Study Analysis

Abstract This chapter presents good practices in the regeneration/rehabilitation and enhancement of small villages, illustrating how interventions shape spaces, communities, and local relationships. It maps diverse Non-European and European case studies, distinguishing between public and private initiatives and evaluating their innovativeness, economic impact, and scale of application. Detailed case sheets provide qualitative analyses and highlight replicable strategies for small-town regeneration. The chapter offers an initial guide of innovative approaches, emphasizing territorial-scale interventions and exemplary cases in Italy, forming a practical framework for sustainable revitalization of inner rural areas.

Within the broad panorama of smart villages, only a few particularly significant cases with different thematic focuses were selected to diversify development strategies. Experiences not cited are nonetheless available in the tables at the end of each section, allowing for a broader understanding of the phenomena addressed. For non-European cases, the research drew not only on scientific literature and books but also on newspaper articles and other sources, aiming to explore a wide range of examples, including niche realities related to Smart Villages. Although most studies focus on European contexts, particular attention was given to projects in less-explored regions of the world, thus expanding the reference framework and enabling the consideration of innovative and sustainable experiences from non-European countries, offering a global and diversified perspective on the phenomenon. Regarding European Smart Villages, the research focused on villages participating in the Smart Rural 21–27 program, aiming to identify the specific vocations of territories and to develop strategies addressing socio-economic stagnation. Given the lack of consolidated academic literature on Italian villages, the study relied extensively on journalistic and non-academic sources, while active participation in association networks facilitated information gathering through interviews and internal discussions with members. Additionally, promotional materials, including blog articles, were produced to systematize and communicate the proposed smart strategies for local territories. This approach provided a comprehensive overview of European cases, with particular attention to detailed examples from Spain and Italy.

L. Lombardo, *Smart Villages*,
SpringerBriefs in Applied Sciences and Technology,
https://doi.org/10.1007/978-3-032-16031-7_2

For the cataloging and mapping of cases, a preliminary verification of each practice's requirements was conducted, analyzing also experiences not directly related to the territory or representing distant dynamics. The first level of verification, defined as "eligibility criteria," included three main components: recognition of the municipality as SMART at the national, European, or international level through the implementation of innovative solutions and media coverage; demographic size, ideally limited to municipalities with no more than 10,000 inhabitants, in line with the definition of small municipalities and considering the fragility of internal areas; and geographical classification, including internal municipalities or, in some cases, coastal municipalities sharing characteristics of isolation and low innovation. The second level, defined as "qualification criteria," comprised eight categories of intervention, including actions related to tourism, commerce, culture, social initiatives, digitalization, heritage conservation, governance, and accessibility, with the requirement that each experience meet at least one of these categories. After verifying first- and second-level requirements, approximately 30 practices worldwide were examined and selected through a qualitative approach based on the highest degree of compliance with the identified criteria. From these, the most representative cases were chosen for in-depth analysis, aiming to define the impacts and values generated and to produce replicable models in different contexts. The final selection prioritized good practices for each type of project action, thereby building a reference framework of exemplary cases, totaling eight experiences.

2.1 Non-European Smart Villages

2.1.1 Anuhi—China

The sustainability of rural areas is facing numerous challenges in China [1]. The Chinese government has chosen the construction of smart villages as an important strategy to achieve sustainable rural development. The smart village is a model that makes full use of the solutions provided by Information and Communication Technologies (ICT) to promote the sustainable development of villages, based on a clear understanding of rural characteristics and development needs [2]. The results of the study showed that the construction and development of smart villages in disadvantaged rural areas is the right choice for sustainable rural development, given the current situation in China. The role of the Chinese government [3] in promoting the construction of smart villages has been to advance the strategic subsystem of the smart village system, to enhance the development of intelligent rural areas through a unified master plan and corresponding support policies. The top-down model of smart village construction currently adopted in China is determined by the country's political and economic system, which is mainly reflected in the centralization of power and the predominance of the public economy. Anhui is a contrast between old and new—the past, present, and future of China. A microcosm of the country's innovative future coexists with programs aimed at protecting and revitalizing centuries-old craftsmanship, culture, and architecture. Anhui is dotted with ancient buildings and communities that have been beautifully preserved and restored, including the village of Hongcun and Tunxi Old Street, located within the mountainous region of Huangshan. According to the plan, several distinctive brands of rural e-commerce products will be cultivated; Internet culture will flourish in rural areas; and the rural digital governance system will be continuously improved. To accelerate the development of digital villages, China will further enhance new infrastructure in rural regions, tap into the potential of new forms and business models, and promote the establishment of a rural digital governance system. China will not only step up the development of facilities related to 5G, Artificial Intelligence (AI), and the Internet of Things (IoT) in rural areas to achieve a deep integration of digital technology and agriculture but will also strive to improve the digital literacy of rural residents.

Geographical Data

Location: Anhui

Country: China

Region: East China

Language: Chinese

Time Zone: GMT + 7

Altitude: 23 m above sea level

Coordinates: 31°52′N 117°17′E

Number of villages involved: 117 villages.

Project Details and Objectives

Number of projects: 117 pilot projects

Project period: 2022—present.

The *Anhui Smart Village Project* was launched by the Anhui provincial government in China in 2017, with the goal of building 100 smart villages across the province by 2020. The objectives of the Anhui Smart Village Project are as follows:

- Improve the quality of life of village residents;
- Increase agricultural efficiency and productivity;
- Attract private investment to rural villages;
- Create employment opportunities in rural areas.

Key Systems Implemented Include

- smart irrigation systems: to control crop irrigation, reducing water and pesticide use;
- crop health monitoring systems: to monitor crop conditions, identify problems early, and take preventive measures;
- waste management systems: to monitor waste generation and disposal, reducing environmental impact;
- security systems: to detect and prevent crimes;
- transportation systems: to improve the mobility of village residents.

Key Themes: Smart agriculture, Rural e-commerce, Shared gardens, Smart healthcare, Poverty reduction, Information and Communication Technologies (ICT).

2.1.2 Banjuwangy—India

The objective of this research is to analyse the implementation of the Smart Village (Kampung) Program as a public service innovation in Banyuwangi Regency [4]. The study contributes to the advancement of public administration science and provides references for the government, academia, citizens, private entities, and stakeholders in developing better and more sustainable public service innovations. The research adopts a qualitative approach, involving research activities aimed at understanding, explaining, and describing social and public phenomena. The realization of public service innovation takes place through the development of villages integrated with information and communication technology based on fiber optics, emphasizing productive and creative economic activities, as well as improvements in education and healthcare, and the eradication of poverty [5]. The results of the study showed that the government of Banyuwangi Regency has successfully implemented the Smart Village (Kampung) Program as an innovation in public services. Banyuwangi Regency won the 2019 Innovative Government Award (IGA) in the "Innovative Regency" category [6]. The Smart Kampung Program is a development concept for newly emerging villages. Its application in Banyuwangi Regency has successfully brought this prestigious recognition to the region. Therefore, this program has been selected as the focus of the research. The establishment of the Smart Village (Kampung) Program can inspire other regions (regencies or cities) to use Banyuwangi's success story as a model and source of learning. Eight priorities form the foundation of the Smart Village (Kampung) Program in Banyuwangi Regency. These priorities include the economy, poverty reduction, education and culture, arts, healthcare, public services, public information, finance, and human resources. Banyuwangi Regency pays great attention to these priorities and dedicates varying degrees of effort to achieving them.

Geographical Data

Location: Banyuwangi Regency

Districts: 24 Kecamatan (districts)

Country: Indonesia

Region: East Java

Language: Javanese

Time Zone: GMT + 7

Altitude: 24–174 m above sea level

Population: 1,540,000

Number of villages involved: 217 villages

Partners and companies: 90

Project period: 2016–present.

Project Details and Objectives

- improve quality of life: using innovative technologies. For example, smart villages can apply artificial intelligence to enhance transportation systems, provide healthcare services, and improve public safety;
- create employment opportunities: through the recruitment of software developers, engineers, and cybersecurity experts;
- attract investment: by encouraging contributions from companies and private investors;
- enhance competitiveness: by strengthening the competitiveness of countries and regions, helping businesses reduce costs, improve efficiency, and increase productivity.

Key Themes: Human resources, Structure and infrastructure, Socialization.

2.1.3 *Blueprint—Niger*

Niger is a landlocked country in West Africa. It borders Algeria to the north, Chad to the east, Nigeria and Benin to the south, and Mali and Burkina Faso to the west. It covers an area of 1,267,000 km^2 (489,700 miles^2), making it the 42nd largest country in the world. The capital and largest city of Niger is Niamey [7].

Niger's population is approximately 23.2 million people (2022), ranking it as the 30th most populous country in the world. The official language of Niger is French, while Hausa, Zarma, and Songhai are the most widely spoken local languages. The main religion is Islam (55%), followed by Christianity (40%), and other religions

(5%). Niger's economy is based on agriculture, livestock, and mining. Its main agricultural products include cotton, sorghum, maize, and millet, while key minerals include uranium, copper, zinc, and gold. Niger remains a poor country, characterized by high rates of illiteracy and infant mortality. Niger is a member of the United Nations (UN), the African Union (AU), and the Economic Community of West African States (ECOWAS). The *Smart Village Blueprint*, piloted in Niger [8], is a practical framework for creating smart villages. It contributes to the implementation of the Niger 2.0 Smart Villages Project, which aims to provide broadband infrastructure to improve Internet access in rural and remote areas of the country, modernize payment systems for farmers, and create digital platforms for agricultural workers. The project adopts a local, integrated, and holistic approach to ensure access to digital services for all citizens in rural communities, integrating digital technologies as a crucial factor for equity and for the qualitative and efficient delivery of SDG-related services for all. Most countries in the African and Asia–Pacific regions have a high proportion of people living in rural areas. For instance, in Burundi and Uganda, 88.24% and 84.23% of the population live in rural districts respectively; in Nepal, 81.76% of the population resides in rural zones. In the Asia–Pacific region, Samoa has 80.74% and the Solomon Islands 78.12% of their populations living in rural areas. The Government of Niger will invest 107 million USD to connect approximately 1.4 million people and 2,111 villages to mobile and broadband services over the next six years.

<u>**Geographical Data**</u>

Country: Niger, Africa

Language: French

Time Zone: UTC/GMT + 1:00

Altitude: 1,000 m above sea level

Coordinates: 17°35′48.77″ N; 8°04′58.26″ E

Number of villages involved: 2,111.

<u>**Objectives of the Smart Village Blueprint**</u>

Project budget: 107 million USD

Partners: SDG Lab

Project period: 2021–2026.

- <u>healthcare</u>: deployment of digital health and telemedicine services for remote consultations and efficient healthcare delivery;
- <u>banking</u>: supporting access to digital financial and investment services;
- <u>education</u>: promoting open and distance learning opportunities to develop skills, ensuring equitable access to quality literacy, lifelong learning, and vocational training;

- employment: providing services for the unemployed and improving their employability;
- agriculture: delivering digital agriculture services to support efficient and productive farming practices;
- crime prevention: developing applications and services for local law enforcement to communicate and share information in real time to protect communities.

Key Themes: Capacity building, Digital transformation, Digital infrastructure, broadband, and connectivity, Digital divide and skills development, ITU-D (International Telecommunication Union—Development Sector), ICT (Information and Communication Technologies).

2.1.4 Chattanooga—USA

Chattanooga is a city in the southern United States, the county seat of Hamilton County, Tennessee. It lies at the confluence of the Tennessee and Chickamauga Rivers, approximately 180 km (112 miles) south of Nashville. Chattanooga is a city rich in history and culture. Founded in 1817, it quickly became an important commercial and transportation hub. The city also served as a significant battlefield during the American Civil War. Today, Chattanooga is a modern and vibrant city. It is home to numerous major companies, including Volkswagen, U.S. Xpress, and BlueCross BlueShield of Tennessee. Chattanooga is also a popular tourist destination thanks to its location along the Tennessee River, its many museums, and its parks [9]. Some of Chattanooga's most popular attractions include: The Tennessee Aquarium, the largest aquarium in the southern United States; The Choo Choo, a former railway station transformed into a museum and shopping complex; The National Civil Rights Museum, dedicated to the history of civil rights in the United States; Point Park, a scenic park overlooking the Tennessee River; The Hunter Museum of American Art, featuring a collection of American artworks; and The Chattanooga Zoo, home to over 1,500 animals. Chattanooga has been at the forefront of smart technology adoption in the United States. In 2014, the city launched the "Smart Village" project, a multi-year plan to transform Chattanooga into a more sustainable, efficient, and livable city. The Smart Village project is based on a series of technologies, including sensors, software, and data analytics. These technologies are used to improve public services, reduce costs, enhance safety, and increase citizens' quality of life. Examples of how the "Smart Village" project has been applied in Chattanooga include:

- the use of sensors to monitor traffic and optimize traffic light timing, reducing travel times and greenhouse gas emissions;
- the use of software to predict infrastructure failures and ensure that public services remain available when needed;
- the use of data analytics to identify at-risk neighbourhoods and implement safety improvement programs.

The Smart Village project has had a positive impact on Chattanooga. Traffic congestion has decreased, costs have been reduced, safety has improved, and the overall quality of life for residents has increased.

Geographical Data

State: Tennessee

Region: Southern United States

Language: English

Time Zone: UTC-5

Elevation: 241 m above sea level.

Project Details and Objectives

Project Budget: 100 million USD

Number of Villages Involved: 10

Partners and Companies: City of Chattanooga, Tennessee Valley Authority, Google, Microsoft, Cisco, IBM, Intel, Oracle, SAP

Project Duration: 5 years.

Objectives of the Chattanooga Smart Villages

- Improve public services through advanced technologies;
- Reduce the city's operational costs;
- Enhance safety through data analysis;
- Improve citizens' quality of life;
- Reduce traffic congestion and greenhouse gas emissions;
- Predict and prevent infrastructure failures;
- Identify and improve at-risk neighborhoods.

Key Themes: Transport, Energy, Safety, Education, Sustainability, Efficiency, Quality of life.

2.1.5 *Olopa—Latin America*

This project supports the implementation of climate-smart agriculture (CSA) technologies and practices and community engagement strategies. It ensures that implementation is carried out through participatory approaches and generates evidence to support the scaling up and mainstreaming of CSA through national and regional stakeholders [10]. For example, activities include conducting baseline studies, engaging with local stakeholders, and developing context-specific agroclimatic services for each Climate-Smart Village (CSV) in Latin America using agroclimatic information and communication technologies. The aim of this research is for organized farmers and indigenous communities in Cauca, Colombia, to achieve sustainable food security for themselves and their families by developing local adaptation plans and adopting CSA practices and technologies. Furthermore, the research seeks to enable farmers in Tuma La Dalia (Nicaragua) and Trifinio (Guatemala and Honduras) to implement CSA options, including alternatives such as agroforestry and agroclimatic services. The success of CSVs is used as evidence to implement local adaptation plans in other villages. In 2014, CCAFS Latin America established a CSV in Olopa, Chiquimula, Guatemala. Here, CCAFS, together with its local partner, the Asociación Regional Campesina Ch'orti' (ASORECH), coordinates the implementation of the CSV approach. Other local actors are directly involved in implementation. For example, during a recent visit, it became clear that a group of women from the La Prensa community is leading the implementation of CSA options within the community. Their motivation is that CSA options help local women increase crop productivity and improve food security, thereby enhancing their livelihoods while adapting to climate change. Gathered in a communal hall, the women of La Prensa learned how concepts such as "probability," "temperature," "climate," and "weather" relate to their crops and deepened their understanding during the first workshop on the Participatory Integrated Climate Services for Agriculture (PICSA) methodology.

Geographical Data

Location: Olopa

District: Chiquimula Department

Country: Guatemala

Language: Spanish

Time Zone: GMT-6

Elevation: 1,500 m above sea level.

Project Details and Objectives

Partners and Organizations: CCAFS Latin America Team, International Center for Tropical Agriculture (CIAT), Asociación Regional Campesina Ch'orti' (ASORECH), Comisión de Acción Social Menonita (CASM), Fundación EcoHábitats, Fundación Procuenca Rio Las Piedras

Project Period: 2014–present.

Objectives

- Implement climate-smart agriculture (CSA) options to improve the yield of local crops;
- Ensure that local families have access to nutritious food through more efficient and sustainable farming practices;
- Educate and equip local communities to address and adapt to the impacts of climate change on agricultural practices;
- Improve the economic and social conditions of local families;

- Educate local communities, especially women, on climate, weather, and probability concepts and their influence on agriculture through workshops and participatory methodologies such as PICSA.

Key Themes: Locally adapted climate variability plans, Agroclimatic information materials, Gender analysis of communities and intervention strategies, Documentation of traditional knowledge integration experiences, Development of improved participatory early warning systems, Formulation of guidelines and planning recommendations to support national-level policy-making.

2.1.6 Sandakpur—Nepal

Sandakpur, a small village in central Nepal, received a new school thanks to a donation from a local humanitarian organization. The school, with a capacity of 100 students, was built from cement and bricks and includes classrooms, bathrooms, a kitchen, and a well. The new school allows children in Sandakpur to receive quality education and have better opportunities for success in life. On April 25, 2015, Nepal was hit by a 7.8 magnitude earthquake, causing significant damage in Sandakpur and destroying many houses and schools. The village population was forced to live in temporary shelters for months. Thanks to humanitarian aid, Sandakpur residents began rebuilding their homes and schools [11]. However, the earthquake left a deep mark on the community, which is still struggling to recover from the tragedy. The village is located in a mountainous area and is often affected by floods and landslides. Residents have started building dams and canals to divert rainwater and prevent flooding. The community is also engaged in tree planting to prevent landslides.

A Smart Village in the hilly region of Nepal, within the Maipokhari and Sulubung villages of rural Sandakpur in eastern Nepal. The village borders India and also has popular tourist sites but lacks basic infrastructure. The project leverages smart resource management, Smart Living, Smart Governance, Smart Village Services, Smart Technology, Smart Tourism, Gender Equity, and Women's Equity, which can support several United Nations Sustainable Development Goals. A five-point Likert scale was used to classify the Smart Village functions, and the Chi-Square test was applied to determine the significance of respondents' opinions on the variables from the collected data. Smart resource utilization is key to a sustainable country, balancing consumption and production using biological and ecological calculators. The survey revealed that most residents depend on natural resources, agriculture, medicinal herbs, and livestock. Tourists are attracted by the pleasant environment and natural beauty. Although agriculture is the primary occupation and the village has potential for the medicinal herbs industry, residents do not own land and basic services such as clean drinking water, irrigation, and ICT-based services are lacking.

Geographical Data

Location: Sandakpur Rural Municipality

District: Ilam, Province No. 1, Nepal

Area: 48.86 km^2 (Ward 1 and 3)

Country: Nepal

Language: Nepali

Time Zone: UTC + 5:45

Elevation: 3,000 m above sea level

Coordinates: 27.01 N, 87.93 E

Population: 1,374.

Project Details and Objectives

Project budget: USD 107 million

Number of villages involved: 2,111

Partners and organizations: CCAFS South Asia Regional Program

Other partners: Nepal Agricultural Research Council (NARC), Ministry of Agricultural Development (MOAD), Local Initiatives for Biodiversity, Research and Development (LI-BIRD), Agriculture and Forestry University (AFU), Nepal

Project period: 2015–present.

Key Themes: Household characteristics, Land use and agriculture, Social infrastructure, Economy, Governance and public services, Environmental concerns, Access to and use of ICT infrastructure, Smart Living, Smart Governance, Smart Village, Smart Services, Smart Technology, Smart Tourism, Gender Equity, Women's Equity.

2.1.7 Smart Farm—South Korea

The Korean approach has been strongly influenced by the European Union (EU) experience, which emphasized the importance of bottom-up territorial development. The Korean government recognizes that agriculture alone is not the main driver of rural employment and wealth creation. Rather, it understands that diversified non-agricultural activities in rural areas are essential to revitalize the rural economy [12].

The main policies relevant to the development of smart rural villages are:

- Creation of a regional innovation system suited to underdeveloped regions;
- Encouraging agriculture to become value-added industries;

- Diversification of rural economic activities and integration of industrial support;
- Improving the well-being of rural residents by enhancing settlement conditions;
- Encouraging rural–urban interaction.

Since the smart rural village initiative as a rural development strategy is closely related to the discussion on rural tourism in Korea, this study also examines past and recent trends in rural tourism strategies pursued by the central government. The transition to digital agriculture in South Korea is inevitable, as the country faces the rapid decline of rural communities. About 46% of the agricultural household population is over 65, indicating not only an aging population but also potential risks to food security. Climate change is another major challenge threatening stable crop production and making it difficult for farmers to predict and respond to weather-related disasters. South Korea is well known as a leader in information and communication technologies (ICT), ranking among the top three countries in the ITU ICT Development Index (IDI) over the past five years. The invention and dissemination of new digital agriculture practices is the new strategy for South Korea to overcome the crisis caused by population aging and climate change. The Smart Farm Innovation Valleys project was launched by the Ministry of Agriculture, Food and Rural Affairs of Korea (MAFRA) in 2019. The project aims to create four Smart Farm Innovation Valleys across the country, each providing a variety of facilities and services to support smart agriculture practices.

Geographical Data

Location: South Korea

Region: East Asia

Language: Korean

Time Zone: GMT + 9

Elevation: 440 m above sea level

Coordinates: 33°0′14.76″ N; 131°9′37.08″ E.

Project Details and Objectives

Project period: 2021–present.

Objectives of the Smart Farm Innovation Valleys project: Promote the adoption of smart agriculture technologies among young farmers, Develop new technologies and practices for smart agriculture, Increase productivity and quality of agricultural products, Improve the sustainability of agriculture, Attract private investment in smart agriculture. The project has already had a positive impact on Korean agriculture, helping to reduce production costs, improve product quality, and increase productivity. It has also contributed to creating new jobs in the agricultural sector. The Smart Farm Innovation Valleys project is ongoing and demonstrates potential to transform agriculture in Korea, making it more efficient, sustainable, and profitable.

Key Themes: ICT, Smart farm.

2.1.8 Uchunju—Azerbaijan

Uchunju Aghali is a village in the Fuzuli District of the Republic of Azerbaijan, with a population of around 500 people. Most residents are farmers, and the village has historically been poor and lacking basic infrastructure. In 2020, the village was destroyed during the Nagorno-Karabakh war. In 2021, the village was rebuilt by the Azerbaijani government [13]. The government had initially explored the idea of "smart villages" before the war: the State Program for Socio-Economic Development of the Regions of the Republic of Azerbaijan 2019–2023 called for the creation of two pilot smart villages. The idea gained momentum after the war, in which Azerbaijan regained more than 8,000 km^2 of territory in and around Nagorno-Karabakh—about 75% of the land lost to Armenian forces in the first war in the 1990s. Over 600,000 Azerbaijanis were displaced during that conflict, many of whom now hope to return. The first smart village project is being implemented in Uchunju Aghali (Zangilan District), focusing on modern technology-based agriculture with joint management and control. The government plans to invest €1.3 billion to build "smart cities and villages." The first smart village project will be implemented in Zangilan District. Preliminary studies have already been conducted. Uchunju Aghali, a village previously destroyed, is the first to be restored. The President has been informed of the project. The project will cover five main components:

- **Housing**—reconstruction and provision of modern, sustainable homes;
- **Production**—implementation of innovative production models to stimulate the local economy, create jobs, and increase productivity;
- **Social Services**—improved access to education and healthcare using smart technologies and digital solutions;
- **Smart Agriculture**—promoting efficient and sustainable agricultural practices while reducing environmental impact;
- **Alternative Energy**—meeting the energy needs of 200 homes using renewable energy sources, in line with the President's vision of creating a "green energy zone."

President Ilham Aliyev has repeatedly emphasized that the liberated territories will become a zone of green energy. This initiative demonstrates the strength and capability of the Azerbaijani state, even amid global economic challenges caused by the pandemic.

Geographical Data

Location: Uchunju Aghali

District: Zangilan

Country: Azerbaijan

Language: Azerbaijani

Time Zone: GMT + 4

Elevation: −30/5,239 m above sea level

Population: 110.

Project Details and Objectives

Project Amount: €1,300,000,000

Project Period: 2021–present.

Project Objectives

- Rebuild destroyed homes and provide modern, sustainable housing using advanced technologies to improve quality of life;
- Implement innovative production models to stimulate the local economy, create jobs, and increase productivity;
- Improve access to social services (education and healthcare) through smart technology and digital solutions;
- Promote smart agriculture practices for efficient, sustainable farming with reduced environmental impact;
- Support the energy needs of 200 homes using alternative energy sources, in line with the vision of creating a "green energy zone."

Key Themes: Social housing, Production, Social services, Smart agriculture, Alternative energy.

2.1.9 Nagoro—Japan

Nagoro is a small mountain village located in the Iya Valley, on the island of Shikoku, Japan. It lies at an elevation of approximately 1,954 m above sea level and currently counts fewer than 30 residents [14]. The village, once home to hundreds of people engaged in forestry and agriculture, has suffered a severe demographic decline due to depopulation and the migration of younger generations to urban centers. The elementary school closed in 2012 after the graduation of its last two students, marking the end of an era for this remote community. The uniqueness of Nagoro lies in its symbolic rebirth as the Village of Dolls—an open-air installation that transforms absence into artistic presence. Around 2003, local resident Tsukimi Ayano, after returning from Osaka to care for her father, was struck by the desolation of her native village. Her first act was simple: she made a scarecrow for her father's field. Soon after, she began sewing life-sized dolls representing the villagers who had left or passed away. Each figure was given a name, a personality, and a story, gradually repopulating the empty spaces of Nagoro—the bus stops, classrooms, fields, and homes. Today, the village hosts around 350 dolls alongside 29 human inhabitants, creating a surreal yet touching tableau that has captured global attention. What began as a personal expression of memory and longing evolved into a powerful example of creative resilience and community identity. Nagoro has become an international tourist attraction, visited by thousands every year, especially during the Kakashi Festival ("Scarecrow Festival") held each October. Media outlets such as National Geographic, BBC, The Guardian, and NHK have reported on this phenomenon, describing it as both poetic and unsettling—a living museum of memory and loss that provokes reflection on the challenges faced by Japan's rural areas. Although not a "smart village" in the technological sense, Nagoro represents a symbolic model of cultural regeneration: a case where art, memory, and identity have reactivated a forgotten place. The project suggests a broader interpretation of "smartness"—one rooted in emotional intelligence, community heritage, and the capacity to reinterpret the void through creativity.

Geographical Data

Location: Nagoro

District: Iya Valley, Tokushima Prefecture

Country: Japan

Language: Japanese

Time Zone: GMT + 9

Elevation: 1,954 m above sea level

Population: 29.

Project Details and Objectives

Project Period: 2003–present.

Key Objectives

- Preserve the memory of former inhabitants through life-sized artistic representations;
- Transform a depopulated village into a cultural and emotional landmark;
- Revitalize the local economy through art-based tourism;
- Encourage reflection on rural depopulation and the value of community;
- Promote creative reuse of abandoned spaces as a form of social and cultural resilience.

Key Themes: Cultural regeneration, art as social memory, creative resilience, rural depopulation, tourism and identity.

2.1.10 Berkhovo—Russia

Bekhovo is a small rural settlement located in the Tula region, approximately 60 km south of Moscow. The village lies on the banks of the Oka River, which serves as a natural border between the Kaluga, Tula, and Moscow regions. With a population of just 58 residents (according to the 2010 census), Bekhovo is part of the Vassily Polenov Fine Arts Museum and National Park, which integrates landscape, art, and heritage into a single cultural system. In 2021, Bekhovo gained international recognition when it was selected by the UN World Tourism Organization (UNWTO) as one of the "Best Tourism Villages" in the world—the only Russian village, and the only one from the former Soviet Union, to receive this distinction. The initiative celebrates rural destinations that embrace tourism as a driver of opportunity, cultural preservation, and sustainable development [15].

Bekhovo's charm lies in its strong connection to Russian art and history. The village's main landmark is the Church of the Life-Giving Trinity, designed in 1906 by Vassily Polenov, a celebrated Russian landscape painter and architect who spent much of his life in the region. His family estate, located just one kilometer away, has been transformed into a museum that continues to attract visitors interested in Russian art, nature, and heritage. Polenov himself was buried in Bekhovo Cemetery, which has become a site of pilgrimage for art enthusiasts and cultural tourists. Today, Bekhovo exemplifies the potential of heritage-based tourism as a sustainable strategy for small rural communities. Through the valorization of its artistic legacy, religious architecture, and natural landscape, the village has managed to transform its isolation into a source of cultural and economic vitality. Its inclusion in the UNWTO program underscores the growing importance of small-scale, community-driven tourism models that balance preservation, identity, and innovation.

Geographical Data

Location: Bekhovo

District: Tula Region

Country: Russia

Language: Russian

Time Zone: GMT + 3

Elevation: 160 m above sea level (approx.)

Population: 58 (2010 census).

Project Details and Objectives

Project Period: 2021–present

Recognition: UNWTO "Best Tourism Villages" (2021).

Project Objectives

- Promote rural tourism as a sustainable development tool for small communities;
- Preserve and valorise the artistic and architectural heritage of Vassily Polenov;
- Strengthen the connection between cultural identity and natural landscape;
- Support local livelihoods through eco- and cultural-tourism initiatives;

- Encourage international recognition of rural destinations through heritage valorisation and soft tourism practices.

Key Themes: Cultural heritage, art-based tourism, landscape preservation, rural identity, sustainable development.

2.1.11 Circa Nundah Village—Australia

Circa Nundah Village [16] is a transit-oriented mixed-use development located in the suburb of Nundah, on the northern fringe of Brisbane, Queensland. The precinct integrates residential, commercial, and retail spaces within a unified public realm, directly interfacing with the Nundah Railway Station and the surrounding community. Developed by Property Solutions and designed by Arkhefield, the project has become one of Brisbane's most recognized urban renewal initiatives. The development includes 186 residential apartments across two towers, a revamped supermarket, retail spaces, and an eight-story commercial tower housing Energex's Northern Metropolitan Office (NMO). The commercial tower achieved a 5-Star Green Star rating, reflecting sustainable building practices, energy efficiency, and high-quality materials. The precinct features interconnected buildings designed to optimize amenity, accessibility, and privacy, while external spaces provide opportunities for outdoor dining and community interaction. Circa Nundah Village exemplifies urban regeneration through mixed-use planning, demonstrating how careful integration of housing, commerce, and public space can revitalize suburban areas. The project is also aligned with the Queensland Government's South East Queensland Regional Plan, which promotes transit-oriented developments to accommodate population growth while preserving environmental and social amenity. Circa Nundah has received multiple awards, including the Urban Renewal Award from UDIA Queensland, highlighting its success as a model for future suburban regeneration.

Geographical Data

Location: Nundah, Brisbane, Queensland

Country: Australia

Language: English

Time Zone: GMT + 10

Elevation: ~20 m above sea level

Population: Integrated mixed-use precinct (186 apartments + commercial tenants).

Project Details and Objectives

Project Period: 2015–present

Recognition: 5-Star Green Star (commercial tower), Urban Renewal Award (UDIA Queensland).

Project Objectives

- Integrate residential, commercial, and retail spaces to create a vibrant community hub;
- Promote transit-oriented development by leveraging proximity to Nundah Railway Station;
- Enhance urban amenity through thoughtfully designed public and outdoor spaces;
- Support sustainable building practices and energy-efficient infrastructure;
- Serve as a model for suburban urban renewal and mixed-use precinct development.

Key Themes: Urban regeneration, mixed-use development, transit-oriented design, sustainable building, community amenity.

2.2 Critical Synthesis of Non-European Smart Villages

1. **Governance and Approach**: Two main governance models emerge: *top-down* and *bottom-up*. China (Anhui) and Azerbaijan (Uchunju) follow a centralized, government-driven approach, where state planning, funding, and technological implementation dominate. Anhui implements 117 smart villages with digital infrastructure, smart irrigation, and e-governance. Uchunju Aghali combines post-conflict reconstruction with smart agriculture and renewable energy. Conversely, South Korea (Smart Farm) and Latin American villages (Olopa, Tuma La Dalia) emphasize community participation, local knowledge, and sustainable agricultural practices, reflecting a bottom-up approach.
2. **Technology and Digitalization**: ICT, digital infrastructure, and smart technologies are central in most projects. China, South Korea, and Niger focus on advanced digital integration (5G, IoT, AI, smart agriculture), while Niger and Nepal incorporate ICT into essential services such as healthcare, education, and employment. Notably, Nagoro (Japan) and Bekhovo (Russia) rely minimally on technology, instead leveraging cultural and heritage-based regeneration, suggesting "smartness" can extend beyond digital innovation.
3. **Key Themes and Priorities**: Projects like Anhui, Banyuwangi (India), and South Korea emphasize economic growth, agricultural productivity, job creation, and investment attraction. Latin American and Nepali initiatives prioritize climate adaptation, resource management, and community resilience, while Nagoro and Bekhovo focus on cultural heritage, tourism, and identity. Niger's Smart Village Blueprint exemplifies a holistic approach, integrating education, healthcare, agriculture, digital services, and financial inclusion.
4. **Socio-Economic and Cultural Impact**: Immediate benefits include improved quality of life, employment, and poverty reduction. Tech-driven villages (China, India, South Korea) deliver tangible economic gains. Heritage- and art-focused

villages (Nagoro, Bekhovo) create cultural and tourism value, showing that smart village strategies can be economic, social, or symbolic.

5. **Challenges and Limitations**: Top-down models may overlook local needs and participation, while bottom-up approaches require strong local governance and longer implementation timelines. Vulnerable regions (Sandakpur, Nepal; Niger) face infrastructure deficits and low digital literacy. Cultural projects offer qualitative benefits that are harder to measure economically.
6. **Comparative Considerations**

Technology and digitalization: China, South Korea, and Niger show systematic ICT integration; Guatemala, Nepal, and India combine ICT with climate and community initiatives.

Governance: Top-down models ensure rapid implementation and large-scale investment but risk low social acceptance; bottom-up models enhance inclusion and sustainability but require more time and local capacity.

Sustainability: Climate and environmental projects (Olopa, Sandakpur) integrate technology and resource management; cultural projects (Nagoro, Bekhovo) expand "smartness" to heritage, identity, and tourism.

Non-European smart villages in Table 2.1 illustrate diverse interpretations of "smartness": from high-tech digitalization to cultural and environmental resilience. Effective strategies balance state support, community engagement, and local resources, combining technology, heritage, and sustainability to adapt to each context.

2.3 European Smart Villages

2.3.1 Alsunga—Latvia

Alsunga is one of the two villages in the second smallest municipality in Latvia, with 1,390 inhabitants [17]. The local community is mainly composed of people with ancient cultural traditions, known as the Suiti. The Suiti Cultural Space was inscribed on the UNESCO Intangible Cultural Heritage List in Need of Urgent Safeguarding in 2009. Alsunga is a good place to live, but ongoing depopulation is reducing available resources, changing the living environment, and increasing the need for services. The main priorities are the development of sustainable local food and a craft market to increase marketing volumes for local producers. Another initiative is the development of the local school. The municipality of Alsunga will be incorporated into a much larger municipality and become a suburban parish. This will require the development of a strong village strategy and a team to maintain Alsunga's development. The entire community benefits from cooperation with residents and entrepreneurs. The smart village is considered a process rather than a static state. Therefore, the "Smart Village

Table 2.1 Non-European smart villages

Non-EU villages	Governance model	Focus/themes	Technology	Budget/Scale	Key impacts
China—Anhui	Top-down, government-led	Smart agriculture, e-commerce, rural digital governance	5G, AI, IoT, digital platforms	117 villages, provincial funding	Productivity, quality of life, investment, rural digital literacy
India—Banyuwangi	Mixed, public innovation	Economy, poverty reduction, education, culture, healthcare	Fiber optics, ICT-enabled services	217 villages, 90 partners	Public service innovation, employment, competitiveness, recognition award
Niger—smart village blueprint	Integrated, holistic, locally adapted	Healthcare, education, agriculture, banking, crime prevention	Broadband, digital platforms	$107m, 2,111 villages	Digital inclusion, SDG implementation, rural connectivity
USA—Chattanooga	Top-down/municipal-led	Transport, energy, safety, education, sustainability	Sensors, data analytics, software	$100m, 10 communities	Traffic reduction, cost savings, safety, quality of life
Guatemala—Olopa	Bottom-up, participatory	Climate-smart agriculture, food security, gender equity	Agroclimatic ICT, participatory tools	Project-based, regional scale	Adaptation to climate change, food security, community capacity building
Nepal—Sandakpur	Bottom-up, humanitarian-led	Resource management, tourism, smart living and governance	ICT for services	$107m, 2,111 villages (regional program)	Post-disaster recovery, livelihood, agriculture, environmental resilience
South Korea—smart farm	Mixed, central gov. with local participation	Smart agriculture, rural economic diversification, youth employment	ICT, digital agriculture tools	National, 4 innovation valleys	Productivity, sustainability, job creation, private investment

(continued)

Table 2.1 (continued)

Non-EU villages	Governance model	Focus/themes	Technology	Budget/Scale	Key impacts
Azerbaijan—Uchunju Aghali	Top-down, post-conflict reconstruction	Housing, smart agriculture, social services, renewable energy	Modern tech-based agriculture, digital solutions	€1.3b, small pilot village	Reconstruction, economic stimulation, sustainability, energy self-sufficiency
Japan—Nagoro	Community-driven, symbolic	Cultural regeneration, tourism, rural depopulation	Minimal ICT	Village-scale	Cultural identity, tourism, creative resilience
Russia—Bekhovo	Community/cultural heritage-focused	Tourism, art-based development, heritage preservation	Minimal ICT	Small village, UNWTO recognition	Tourism, heritage valorization, community identity
Australia—Circa Nundah Village	Private/public partnership	Mixed-use development, urban regeneration, sustainability	Green building, smart urban design	Suburban precinct, 186 apartments	Community hub, sustainable building, transit-oriented development

Action Group" was established as a volunteer association involving representatives of the Municipal Council President, NGOs, local authorities, and residents.

Alsunga Strategy: **Objective 1**: Development of small local value-based businesses; **Objective 2**: Stabilize the village population.

- Weekly market (*Suitu tirgus*) of local food producers and artisans—highly visited and popular among residents from other regions of Latvia;
- Video on Alsunga Smart Village promoting the area as a tourist destination, highlighting local cultural history, traditions, and the production of small local producers and artisans (around 55,000 views);
- Create a more interesting and exploratory environment;
- Mobile application to encourage cooperation among tourism entrepreneurs;
- Augmented or virtual reality learning tools created for the local school environment;
- Co-working space where various independent entrepreneurs can operate in a shared common environment;
- New tourism product—highlighting and emphasizing the local area's cultural history.

Geographical Data

Location: Alsunga

Country: Latvia

Region: Latvia

Language: Latvian

Time zone: EET, UTC + 2

Altitude: 200 m above sea level

Population: 23,438.

Project Details and Objectives

Project budget: €7,700,000

Number of villages involved: 1

Project period: Latvian National Development Plan 2014-2020 (NAP2020).

The specific objectives of the Alsunga Smart Village are:

- Improve residents' quality of life by creating a safer, healthier, and more sustainable environment;
- Develop a more resilient and competitive local economy;
- Create a model for the development of other smart villages in Latvia and other countries.

Key Themes: Digitalization, Strong cultural traditions, High quality of life, Nature as a resource, Good collaboration between the community and local government.

2.3.2 Athens—Greece

Athens, a city that emerged from the unification of smaller centers in Attica during the late Mycenaean period, held a prestigious position as the capital of Attica [18]. During the archaic phase of its history, Athens experienced monarchical rule, common to many other urban centers in Greece. From the eighth century BCE onwards, the absolute power of Athens' monarchs gave way to aristocracies, with governance entrusted to magistrates elected by the city's ruling classes. Initially, the archons had ten-year mandates, but over time this office evolved into a collegial body composed of nine members with annual terms. The growing conflicts between peasants and the landowning aristocracy in the seventh and sixth centuries BCE necessitated a shift toward more inclusive political forms. In 594 BCE, Solon, one of the archons, intervened on behalf of small farmers by introducing a political system based on income classes, aiming to limit aristocratic power. However, social tensions persisted, leading to the tyranny of Peisistratus in 560 BCE, under which Athens experienced a period of peace and prosperity. Although Peisistratus' tyranny was initially beneficial, it degenerated over the course of the sixth century BCE under the rule of his successors, reigniting social conflicts within the Athenian population. ELECTRA ENERGY is an organization that has been working since 2016 for the transition toward a democratic, inclusive, efficient, and sustainable energy system centered on citizens and local communities. Greece is becoming a regional leader in community energy production thanks to new regulatory frameworks introduced in 2018, implementing European legislation. Electra Energy, based in Athens, is a cooperative that promotes energy efficiency and savings through awareness campaigns, the development of new partnerships, and local networks. It seeks to address energy poverty by donating part of its production to the families most in need. Electra Energy aims to build a solar photovoltaic power plant in the Corinth region based on a community "virtual-net-metering" model. This means that the model will allow community members to purchase a share of the installation according to their consumption and benefit from solar production, reducing their annual electricity bill. The solar plant will supply energy to over 130 homes and small businesses, producing 750,000 kWh of clean solar energy per year for 25 years and saving a total of 7,500 tons of greenhouse gas emissions.

Geographical Data

Location: Athens

Country: Greece

Language: Greek

Time zone: CET, UTC + 1 (EEST)

Altitude: 20 m above sea level

Population: 4,013,368.

Project Details and Objectives

Number of villages involved: 1 city

Project period: 2018–present

Objectives: Public schools and photovoltaic systems, By(e) Plastic Project, "Hyperion": the first community solar investment in Greece, powered by citizens, Guide for energy communities.

Key Themes: Energy community.

2.3.3 Kythera—Greece

The island of Kythera is located at the intersection of three seas: the Ionian, the Aegean, and the Cretan Sea. Kythera features a mild Mediterranean climate, humidity, and strong winds, especially in winter [19]. The frequent alternation of landscapes on this mountainous island, with numerous valleys leading to the sea, results in many beautiful beaches. The island has many settlements and a population of 4,000 inhabitants; the villages are numerous, small, and scattered throughout the geographic area. Administratively, the island is organized as a municipality with the community of Antikythera, the nearby homonymous island, under the province of Kythera, which belongs to the region of Attica. The island's economy is based on rural life and tourism, while many locals work in services, trade, and public organizations. Population aging and youth emigration for education and employment opportunities are major challenges for the island. Due to its geographic position, Kythera has limited access to markets, as its nearest coast is two hours by boat, and transportation connections are often restricted. Its unique natural environment is its most important asset; the quality of life, natural beauty, socially aware community, fertile land, recognized local products, and island history are among its key advantages. In the Alsunga-inspired strategy, particular attention is given to enhancing local production, focusing on products such as olive oil, honey, and aromatic and medicinal plants. This approach aims not only to promote and preserve Kythera's landscapes, ecosystems, and biodiversity but also to support quality tourism through alternative forms, highlighting the island's cultural heritage, natural beauty, and unique local products. The local community, characterized by a strong sense of identity supported by various associations and local organizations, actively participates in the Smart Rural 21 project. The goal is to generate new ideas and inspiration for developing an advanced strategic plan for implementing innovative smart village approaches, thus contributing to the sustainable growth and well-being of Kythera.

Geographical Data

Location: Kythera

Country: Greece

Region: Attica

Language: Greek

Time zone: EET, UTC + 2

Altitude: 200 m above sea level

Population: 3,500.

Project Details and Objectives

Project budget: €7,400,000

Number of villages involved: 8 municipalities, 11 associations and groups

Project period: 2014–2020.

Objectives

- Develop small local value-based businesses;
- Stabilize the village population;
- Retain young people on the island;
- Improve agriculture and promote local production, new technologies in agriculture, and smart farming solutions;
- Enhance environmental sustainability;
- Support entrepreneurial activities;
- Increase quality tourism; support alternative tourism and promote the area as a tourist destination; promote it as a safe destination thanks to the existing hospital;
- Improve services (transport, education, and healthcare).

Key Themes: Strengthening cooperation between municipalities, Developing local and regional strategies, Supporting local development actions.

2.3.4 Lavancourt—France

Access'R is a three-year project. It was launched in 2018 by Leader France, the national federation of rural territories, together with 15 French and European partners, all engaged in the European LEADER rural development program. This project was developed and implemented under the collective mobilization call for rural development launched by the National Rural Network. Its purpose is to identify, promote, and share good practices and innovative approaches aimed at improving accessibility, maintaining, and revitalizing services for populations in rural areas.

The image of rural areas was reinforced by the pandemic [20]. However, the preservation and development of public services remain a major challenge for these rural territories. After an initial collection of experiences published in September 2020, Leader France and its partners now offer a practical guide. It provides analytical tools to better understand the levers to facilitate access to services, along with tool sheets describing the steps and mechanisms to act. Maintaining and strengthening public services is an important issue in rural areas. This topic, which is one of the main concerns of residents, requires considering new forms of presence in postal, educational, medical, and social services. ACCESS'R aims to improve the accessibility of rural areas in France, promote connectivity, and foster collaboration among the actors involved in the process. It includes a network of smart villages sharing pathways aimed at connecting small inland towns and experimenting with practices for mutual enhancement.

Key actions: Cooperation and sharing, Mobile services and mobility solutions, SMART and digital approaches, Funding mechanisms.

Geographical Data

Location: Northern France

Country: France

Region: multiple regions

Language: French

Time zone: UTC + 2

Altitude: 372 m above sea level

Population: 336.

Project Details and Objectives

Number of villages involved: 10 French villages and 6 European partners

Project period: 36 months, 2018–2021.

Objectives

- Develop smart solutions to improve mobility, safety, energy efficiency, and sustainability in rural villages;
- Create a model for the dissemination of smart solutions;
- Raise awareness of the benefits of smart solutions for rural villages;
- Smart transportation systems to reduce traffic and improve road safety;
- Smart security systems to detect and prevent crimes;
- Smart energy efficiency systems to reduce energy costs and improve environmental sustainability;
- Smart waste management systems to reduce waste production and improve environmental quality.

Key Themes: Public services, Health, Youth education, Sports and culture, Commerce, Digital mobility.

2.3.5 Penela—Portugal

The Municipality of Penela has a population of approximately 5,983 inhabitants. It is in the Coimbra District, in the Central region of Portugal. Ferraria de São João is a schist village within this municipality, situated in a remote position on the top of a mountain. It has 43 inhabitants, only 9% of whom are under 20 years old. The village offers a high quality of life to its residents, providing access to health, education, culture, and leisure due to its proximity to the town of Penela and the city of Coimbra [21]. The local community, despite aging, plays an active role in the village development decision-making process. The village has experienced increasing investments in tourism and agriculture. Depopulation and aging are the most significant challenges faced by the village and are central to defining a smart strategy. These challenges result in a limited capacity to attract new residents and skilled human resources. Ferraria de São João stands out for the high quality of life it offers, thanks to its natural environment and the community's ability to mobilize volunteers, particularly through tourism-related initiatives. The village has shown growing dynamism in the tourism, agricultural, and forestry sectors, linked to the sustainable use of resources and the recovery and enhancement of local traditions. The strategy focuses on enhancing the natural heritage of the village and surrounding territory and on promoting its rural and community identity, as well as intergenerational dialogue based on traditional values. It is primarily aimed at creating new cultural, social, and entertainment activities that encourage integration between residents and tourists, strengthening social and territorial cohesion and reinforcing social and cultural dynamics. It focuses on improving the built environment through the redevelopment of public spaces and facilities, providing necessary conditions for the population, and creating the framework to attract and retain new residents and investors. The strategy also emphasizes the promotion of local identity through the recovery of village assets and intangible heritage, considered a fundamental resource for local economic development. Additionally, it aims to enhance and support investments in the local economy, contributing to the sustainable development of the village, the well-being of the population, reversing depopulation trends, and fostering regional development.

Geographical Data

Location: Penela

Country: Portugal

Region: Coimbra District

Language: Portuguese

Time zone: Azores, Portugal (GMT-1)

Altitude: 164 m above sea level

Population: 6,594.

Project Details and Objectives

Project budget: €5,004,130.20

Number of villages involved: 2 villages

Project period: 2020–2029.

Objectives

- Enhance the natural heritage of the village and surrounding territory;
- Promote the village's rural and community identity;
- Encourage intergenerational dialogue based on traditional values;
- Create new cultural, social, and entertainment activities;
- Strengthen social and territorial cohesion;
- Redevelop public spaces and facilities;
- Recover village assets and intangible heritage to enhance local identity;
- Support investments in the local economy;
- Improve population well-being and counter depopulation;
- Contribute to regional development.

Key Themes: Conservation of natural heritage, Promotion of environmental sustainability, Promotion of protection and safety of people and assets.

2.3.6 Phini—Cyprus

Located in the Troödos Mountains of Cyprus, Phini is a village with a long history. Its origins likely date back to antiquity, when the geographic location and natural resources of the area supported the development of small agricultural communities. During the Byzantine period, Phini was influenced by the cultural and religious changes on the island, possibly including the construction of monasteries and churches. Under Venetian rule, the village participated in the economic expansion and commercial exchanges that characterized the island during that time [22]. The transition to Ottoman rule marked a change, with less favorable economic conditions, but maintained a degree of stability. In the twentieth century, Phini experienced the historical events leading to Cyprus's independence in 1960, following the fate of the entire island. The village preserves examples of traditional architecture, with stone buildings and structures typical of Cypriot mountain communities. The local economy is primarily based on agriculture and livestock, which continue to play an important role in village life. The village's traditions, closely tied to agricultural cycles and religious festivals, reflect the community's cultural heritage, which has preserved many of its ancient customs over time. Since 2019, the Phini Community Council has organized an annual National Cultural Conference, each year dedicated to a specific theme. In 2019, the first conference focused on "Ceramics of Yesterday and Today in Phini," while in 2020 the theme was "Phini: Historical and Cultural Heritage from Antiquity to Today." In 2021, the third conference focused on "Youth Life and Retention in Mountainous Areas and Their Needs." Despite pandemic restrictions, more than 180 delegates participate on average each year. The conferences provide space for local product exhibitions, such as ceramics, handicrafts, and the famous local sweet "Phiniotiko Loukomi." These events allow participants, especially young people, to share ideas and proposals regarding policies on the discussed themes, with particular attention to funding opportunities for investments in mountain areas, agriculture, handicrafts, and tourism. The organization involves various local and regional actors, including schools, NGOs, government bodies, and national rural networks. A bulletin summarizing the recommendations emerging from the conference is also produced and distributed to key national entities.

Geographical Data

Location: Troödos Mountains, Cyprus

Country: Cyprus

Language: Greek

Population: 391 people

Time zone: (CET), UTC + 3 (EEST)

Altitude: 1,000 m above sea level

Number of inhabitants: 3,604.

Project Details and Objectives

Number of villages involved: 1 community

Project period: 2019–present.

Objectives

- Promote local cultural heritage: Enhance the history, traditions, and artisanal activities, with particular attention to ceramics;
- Engage the community and youth: Express ideas and aspirations for the future, encouraging active participation in community life;
- Develop policies for mountain areas: Discuss and propose policies to improve living conditions (education, healthcare, transport, employment, and basic services);
- Promote investment opportunities: Provide funding opportunities for agriculture, tourism, handicrafts, and services, supporting local economic development;
- Facilitate cooperation among local and national actors: Strengthen collaborations among local authorities, schools, NGOs, and others.

Key Themes: Infrastructure and connectivity, Education and training, Environmental and energy sustainability, Economic development and tourism, Culture and traditions.

2.3.7 *Raudanmaa—Finland*

Raudanmaa is a rural area with approximately 600 inhabitants, composed of several small villages. People live very close to nature in this area, which is a major advantage that our villages can offer both new residents and visitors. Our main aspiration is to be an innovative, service-oriented village where daily life runs smoothly. We aim to introduce new enabling innovations to revitalize the village economy and improve living conditions. We are proud of our local community, which is very active and supportive of new initiatives. Several associations are active, and regular meetings are organized among villagers on issues affecting everyone. We have built our water supply, sewage system, and fiber-optic network. One of our main current priorities is to create a well-functioning multi-service smart town hall that provides activities and services to residents and visitors. We have collected the opinions and ideas of all villagers to develop a plan for the town hall and are currently developing detailed plans for its construction, operation, and sustainability. For Raudanmaa [23], "smartness" means being an innovative, service-oriented village where daily life proceeds smoothly. The goal is to keep the rural village area vibrant while preserving its distinctive physical and cultural environment, as well as the heritage created by the common, traditional lifestyle of the villages. New enabling innovations are introduced to revitalize the village economy and improve living conditions. Fast network connections and teleworking opportunities provide new possibilities, which are already partly in use and could be exploited and commercialized in innovative ways. Digital services can facilitate effective cooperation and open information sharing both within the community and with visitors and external organizations. Vital sources of life, social and physical well-being are supported by providing innovative services to all villagers in this rural area. The main development aspirations for smart village activities are linked to three key areas: environment, culture, and vitality. "Smart" naturally means the use of new technology, but social intelligence is equally crucial as we seek to enhance the traditional village culture and bring it to a new level in the digital context.

Geographical Data

Location: Raudanmaa

Country: Finland

Region: Satakunta

Language: Finnish

Time zone: UTC + 03:00

Altitude: 121 m above sea level

Number of inhabitants: 510.

Project Details and Objectives

Number of villages involved: 1 village

Project period: 2017–2019

Own projects: Kautilya Water Cooperative 2010, Raudanmaa Fiber-Optic Cooperative 2017, Smart Town Hall Planning Project 07/2019–05/2021, Raudanmaa Village Cooperative 2020.

Objectives

- Promote innovation and provide service-oriented initiatives that improve residents' daily quality of life;
- Keep the rural village area vibrant while preserving its distinctive physical and cultural environment;
- Introduce new enabling factors to revitalize the local economy;
- Leverage fast network connections, teleworking opportunities, and digital services to create new economic opportunities.

Key Themes: Enhancing village vitality, Quality of rural life and clean environment, Innovative village culture.

2.3.8 Remetea—Romania

Located in the mountainous area of the Giurgiu Basin in Harghita County, Romania, Remetea is one of the largest villages in the region. The village lies at the foot of the Giurgiu volcanic mountains at an average altitude of 750 m above sea level. The population is 6,171 inhabitants, the vast majority being of Hungarian ethnicity and Roman Catholic faith [24]. The total area of the village is 10,774 ha, of which the village itself covers 1,145 ha. Remetea and the surrounding villages typically experience long cold winters and short summers. The average annual temperature ranges between 5.1 and 5.5 °C. The Mureş River, which enters Remetea from the south and forms large curves dividing the village in two, is the largest local river. On the right bank of the Mureş lies the Nagyrét swamp, home to rare plant species that have survived since the Ice Age. In 1980, four hectares of this land were declared a nature reserve. Administrative management of the village is ensured by the local council as a deliberative authority, composed of 15 local councillors acting as the executive authority. The specialist structure of the municipality is a small-scale organization that executes the council's decisions and the mayor's directives, aiming to respond to all community needs. The "smart village" concept addresses emerging needs identified within the community through the following actions: expanding employment opportunities for both young and elderly residents by promoting remote and online work; diversifying the job market by encouraging ICT start-ups; promoting digital marketing; organizing online volunteering and charity activities; using online platforms for intergenerational knowledge transfer; providing online education for children, youth, and adults; creating an educational platform for schools; offering after-school programs focused on ICT skills; developing a presentation platform with useful information on health, programming, preliminary information on services, and medical advice; organizing digital campaigns on healthy lifestyles; monitoring soil and plant conditions and collecting data via sensors; consumer education; creating an app for water and sewage services called "RemeteApp"; digitalizing administration, maintaining registries and genealogical records; live streaming of events; mobile application surveys; and promoting public engagement.

Geographical Data

Location: Remetea

Country: Romania

Region: Transylvania

Language: Romanian

Time zone: (EET), UTC + 2

Altitude: 750 m above sea level

Number of inhabitants: 6,261.

Project Details and Objectives

Number of villages involved: 1 village

Project period: by 2021.

Objectives

- Address youth emigration;
- Diversify educational opportunities;
- Open non-formal online learning opportunities for adults;
- Improve healthcare services and prevention;
- Make public services more efficient through online service development;
- Environmental management;
- Provide reliable and authentic information to residents.

Key Themes: Community and local identity, E-governance and accessible local services, Living standards and environmental issues, Digital education and SMART students, Limited entrepreneurial activity and lack of market access.

2.3.9 *Zaragoza—Spain*

Zaragoza, with over 2,000 years of history, is renowned for its imposing Basilica del Pilar, its Roman origins, rich cultural diversity, and the hospitality of its people. Located on the banks of the Ebro River, it is the capital of Aragón and the fifth largest city in Spain by population. A crucial hub connecting major routes to other large Spanish cities, Zaragoza is an ideal destination for a city break [25]. The international airport, high-speed train station, and well-developed road network allow reaching major destinations in northern and central Spain, as well as southern France, within 2–3 h. Zaragoza is famous for its rich historical and monumental heritage, reflecting the influences of four ancient cultures. The Basilica of Nuestra Señora del Pilar, the city's symbol and the second cathedral of Zaragoza, houses majestic works by Francisco de Goya and the revered sculpture of the Virgen del Pilar. Nearby is the Cathedral of El Salvador, the city's main historical temple and an emblematic example of Aragonese Mudéjar art, recognized as a UNESCO World Heritage Site. The cathedral also preserves one of the most important collections of Flemish tapestries from the fifteenth–seventeenth centuries. The main actors are families, shops, businesses, associations, and institutions located within 500 m of the photovoltaic installation. Barrio Solar is an initiative aimed at reaching all residents in the neighborhood where it is installed. A portion of the energy it generates is allocated to households in energy poverty, without any monthly fee, allowing them to benefit directly from the savings on their electricity bill generated by solar self-consumption. Both EDP customers and any other company can participate in Barrio Solar. The project, coordinated by ECODES and implemented in collaboration with Edp Solar, Zaragoza City Council, Schneider Electric Foundation, and Fundaciòn edp, involves the installation of a 100 kWp photovoltaic system on the roof of a public building. By paying a monthly subscription (around €7), residents within 500 m of the installation can benefit from a savings of approximately 30% on energy costs. The subscription also covers 10% of vulnerable households living in energy poverty in the neighborhood, who can access the benefits at no cost. The project, currently under implementation, is based on strong citizen information, education, and engagement activities.

Geographical Data

Location: Zaragoza

Country: Spain

Region: Aragón

Language: Spanish

Time zone: (EET), UTC + 2 (CEST)

Altitude: 208 m above sea level

Number of inhabitants: 682,513.

Project Details and Objectives

Number of villages involved: Entire city

Project period: by 2021.

Objectives

- Provide sustainable and accessible energy;
- Combat energy poverty;
- Promote social and economic inclusion;
- Foster community participation;
- Encourage multi-stakeholder collaboration.

Key Themes: Energy community.

2.3.10 Stanz—Austria

Stanz im Mürztal is in a rural area in the Austrian province of Styria, with a total population of 1,855 inhabitants. Stanz is a local authority with elected political representatives. In recent years, the village has successfully developed initiatives focused on revitalizing the town center, creating new mobility options, and implementing measures to improve the quality of life. Residents value the quality of life in the village: the results of a recent survey revealed that the school, kindergarten, and civic life received the highest ratings. The region is industrial and has been affected by structural economic changes over the past decades, which has contributed to strong depopulation and an aging population [26]. The small local economy is reflected in the high percentage of residents commuting to work in nearby areas. Stanz has decided to take new approaches to community building: strengthening direct democracy, actively involving citizens in municipal development, constructing sustainable infrastructure, and improving local resilience, recognizing that people and their engagement are key resources. An additional advantage is that 80% of Stanz's municipal area is forested, offering opportunities for sustainable energy production. Stanz introduced a new local food supplier that ensures 50% of goods come directly from local producers. To reduce waste, a nearby restaurant uses products from the store—such as fruits, vegetables, and meat—cooking and offering them as take-away food in the store. Additionally, a vending machine provides a 24/7 food service. To promote alternative and sustainable mobility, Stanz launched the "Mitfahr-Bankerl" initiative—benches painted yellow/green placed along public roads, each with an umbrella attached. When the umbrella is open, it indicates that the person sitting on the bench is looking for a ride. Over the past three years, vacant buildings in the village center have been sustainably renovated, including wooden construction elements. This led to the creation of a new multifunctional center that includes apartments for young and elderly residents, the store, two service facilities, and the new municipal administrative building, thereby improving community administration.

Geographical Data

Location: Stanz

Country: Austria

Language: German

Time zone: (CET), UTC + 1

Altitude: 1,040 m above sea level

Number of inhabitants: 441.

Project Details and Objectives

Number of villages involved: 1 village

Project period: until 2021.

Objectives

- Achieve energy self-sufficiency by 2030;
- Establish sustainable tourism as a new economic pillar;
- Further develop civil society.

Key Themes: Population decline, Depopulation of the town centre, Lack of employment, Loss of supply infrastructure, Lack of local shops and housing, Dependence on car traffic, Lack of participation in decision-making processes.

2.3.11 Tomaszyn—Holland

Tomaszyn is a small village in Warmia and Masuria, Poland. It consists of ten families and one cooperative. It is considered the "green lung" of Poland and has the lowest population density in the country. The voivodeship hosts eight landscape parks, home to natural phenomena of European significance. Tomaszyn was an agricultural and prosperous village for many centuries, with the first written records dating back to 1410. However, most of the village buildings were destroyed during World War II. The population decreased to four farms and a few vacationers, reaching its lowest point in 2017 [27]. In 2018, the Ostoja Natury Cooperative was established in Tomaszyn. Ostoja Natury is the "agricultural ecosystem of tomorrow," which serves as the reference model for a smart village farm. This organic farm operates as a closed-loop system, where waste becomes fuel, produces energy, allows year-round food production, and operates efficiently to ensure higher yields. Similar to a prehistoric forest, everything is part of a system, and nothing is wasted. The goal is to create an ecosystem designed to provide economic efficiency while minimizing environmental impact. Traditional methods honed over centuries are renewed and implemented using modern technological innovations. The Ostoja Natury program—Ostoja Wioska 3.0—involves preparing a complete “know-how” covering all key aspects of running an organic farm, including:

- Production methods adapted to local pedoclimatic conditions;
- A set of technological solutions;
- Optimal machinery park offerings;
- Robotics and automation in agricultural production and processing;
- Building a modern distribution platform enabling direct customer interaction and sales;
- Access to a seed bank (BIO seeds);
- Construction of a model, self-sufficient, waste-free habitat for rural farms;
- Implementation and optimization of a functional business model;
- Formation of an independent producers’ and cooperatives group based on shared distribution;
- Platforms for communication, technology, knowledge, and seeds.

Geographical Data

Location: Tomaszyn

Country: Poland

Region: Gmina Olsztynek

Language: Polish

Time zone: UTC + 01:00

Altitude: 135 m above sea level

Number of inhabitants: 18.

Project Details and Objectives

Project budget: €7,700,000

Number of villages involved: 1

Project period: 21 months.

Objectives

- Improve the quality of life of residents;
- Reduce the village’s environmental impact;
- Make the village more energy-efficient;
- Create a safer and more welcoming environment for residents;
- Develop a more resilient local economy.

Key Themes: Ostoja Natury Village 3.0, Organic farming center, Waste-free, self-sufficient habitat for rural families, Passive rain irrigation system, Passive greenhouses.

2.3.12 Vuollerim—Sweden

Vuollerim is a village located in the municipality of Jokkmokk, in Norrbotten County, northern Sweden, situated in the heart of Lapland. Its history is closely linked to the presence of the Sámi people, the indigenous inhabitants of the region, known for reindeer herding and a lifestyle deeply connected to nature [28]. The village has likely been inhabited for centuries, following the typical dynamics and transformations of the Lapland region. Over time, it has experienced significant changes influenced by industrialization and urban migration, facing the challenge of maintaining its population and cultural identity. In recent decades, tourism has contributed to revitalizing the local economy, with particular emphasis on cultural and natural experiences. Visitors are attracted by the opportunity to learn about Sámi culture, participate in traditional activities such as reindeer excursions, ice fishing, and taste local cuisine, all while immersed in an unspoiled landscape. A major event is the "1000 Ice Lanterns Festival," held every February, featuring thousands of ice lanterns that create a magical atmosphere celebrating winter and the beauty of the Lapland landscape. Vuollerim's "smart" strategy is based on an innovative and collaborative local economic development model, founded on active community participation. The village hosts approximately 60 small businesses, 40 non-profit associations, and 7 rural enterprises, all working in synergy to promote sustainable territorial growth. A key role in this strategy is played by Vuollerimbygdens Ekonomiska Förening, an economic association that initiated a new approach to collective management of local enterprises. Over the past 20 years, this model has spread, involving various local actors. The goal is to optimize resources, foster employment, and improve residents' quality of life, creating a resilient and sustainable local economy based on cooperation and innovation. By supporting small businesses and local entrepreneurial initiatives, Vuollerim addresses common rural challenges such as depopulation and aging populations, promoting an environment conducive to social innovation and economic development.

Geographical Data

Location: Vuollerim

Country: Sweden

Region: Norrbotten (Norrbottens län)

Language: Swedish

Time zone: CET (Central European Time), UTC + 1; CEST (Central European Summer Time), UTC + 2

Altitude: 400–500 m above sea level

Population: 732.

Project Details and Objectives

Number of villages involved: Entire community

Project duration: Over 20 years.

Objectives

- Promote local economic development;
- Foster cooperation between businesses and the community;
- Create new opportunities for youth and families;
- Support innovation and resilience in rural areas;
- Improve quality of life through active community engagement.

Key Themes: Community cooperatives, Local economic development, Community cooperation, Opportunities for youth and families, Rural innovation, Quality of life and community engagement.

2.3.13 Teruel—Spain

Teruel, located on a hill in southern Aragón, preserves an important medieval heritage in its historic center. The cathedral tower and those of San Salvador, San Martín, and San Pedro are jewels of a monumental complex in Mudejar style, recognized as a UNESCO World Heritage Site [29]. The city, legendary due to the famous story of the Lovers of Teruel, retains charming corners such as the central Plaza del Torico, the aqueduct with its arches, and the famous Óvalo avenue. Since ancient times, various prehistoric and Iberian civilizations settled near present-day Teruel. However, the city reached its peak during the Middle Ages, when its perimeter was fortified, bordering the Muslim Kingdom of Valencia. Islamic influence left Teruel and Aragón with the most important Spanish Mudejar architectural complex, combining Gothic elements with Muslim architectural features. The cityscape includes other remarkable Mudejar towers, such as the tower of San Salvador on Nueva Street, with its brick façade, double arch orders, and rich ceramic decorations. The Santa María de Albarracín Foundation, founded in 1996, represents an excellent example of innovative cultural heritage management, focusing on the restoration and enhancement of Albarracín's traditional plaster architecture. Located in the old Episcopal Palace of the city, restored through applied training programs, the Foundation is committed to promoting Albarracín as a prominent cultural emblem. The initiative primarily focuses on plaster, a distinctive material in local architecture, encouraging the compatible restoration of traditional structures. Through courses, seminars, exhibitions, and concerts, the Foundation runs a dynamic cultural program involving both residents and visitors. Of particular significance is the support for artistic creation and training of movable heritage restorers, contributing to revitalizing Albarracín and positioning it as a leading cultural reference nationally and internationally.

Geographical Data

Location: Valencia

District: Castellón

Country: Spain

Language: Spanish—Valencian

Time zone: UTC/GMT + 1

Altitude: 915 m above sea level

Population: 35,675.

Project Details and Objectives

- Socio-cultural and economic development of the territory;
- Restoration, conservation, and management of cultural assets;
- Revitalization and promotion of cultural and historical heritage;
- Promotion of diverse cultural and educational activities;
- Establishment of collaborative modalities among different entities;
- Advisory services on maintenance and restoration techniques;
- Artisan training;
- Organization of events and awareness campaigns;
- Research and documentation;
- Restoration of historical plaster buildings;
- Creation of new public and cultural spaces;
- Promotion of sustainable tourism.

Themes Addressed: Restoration of traditional plaster architecture, Strategies and pathways for revitalizing ancient construction techniques.

2.3.14 Civita di Bagnoregio—Italy

Civita is a hamlet of the municipality of Bagnoregio, in the province of Viterbo, Lazio, and is part of "Italy's most beautiful villages." It is famous as "The Dying

City." Inhabited by about ten people and located in an isolated position, it is accessible only via a reinforced concrete pedestrian bridge built in 1965. The bridge can be crossed on foot, but recently the municipality of Bagnoregio allowed it to be used by bicycles and motorcycles. Its isolation is due to the progressive erosion of the hill and surrounding valley, giving rise to the nickname "the dying town." Civita is a small jewel of Etruscan architecture, immersed in a unique atmosphere, with a cluster of houses enjoying a spectacular natural view from their privileged position. Situated near Lake Bolsena and the Umbria border, Civita di Bagnoregio is one of the oldest and most beautiful villages in Italy. Its fragility and slow decline contribute to its charm, making it a must-visit urban masterpiece. Civita di Bagnoregio is at the center of an agreement between the association founded by Gianfranco Imperatori and the IUAV University of Venice [30] to create a new project for the safeguarding of Italian architectural, historical, and landscape heritage. The initiative attracts professors, international experts, students, and intellectuals. It includes courses and seminars for students and graduates, as well as roundtable discussions with specialists in architecture, landscape, and sociology from Italy and Europe. "The dying town," as described by the writer Bonaventura Tecchi, aims to become a model for landscape conservation in Italy, representing the ideal Italian village. It aspires to integrate the care of beauty, history, and identity with the cultural and economic development of the local community, pursuing a sustainable future. The project has secured €2 million for hydrogeological risk mitigation through a program agreement between the Ministry of the Environment and the Lazio Region, with an additional €850,000 contribution from the Italian Prime Minister's Office.

Geographical Data

Location: Civita di Bagnoregio

Country: Italy

Region: Lazio

Language: Italian

Time zone: Rome (CET)

Altitude: 200 m above sea level

Population: 3,303.

Project Details and Objectives

Project funding: €2,850,000

Number of villages involved: 1

Project period: 2011–present.

The objective of the agreement between the association and IUAV University is to create a new project dedicated to the protection of Italian architectural, historical, and landscape heritage. It attracts professors, international experts, students, and intellectuals through courses, seminars, and roundtables with specialists in architecture, landscape, and sociology. Known as "the dying town" according to Bonaventura Tecchi, Civita aims to become a model of landscape conservation, symbolizing the ideal Italian village. The project integrates the care of historical and cultural heritage with the cultural and economic development of the local community, aspiring to a sustainable future.

Key Themes: Enhancement and enjoyment of architectural heritage, Workshops and educational activities, Inventory and documentation of architectural features, Tourism promotion, Film production initiatives, Mitigation of hydrogeological risk and accessibility improvements.

2.3.15 Craco—Italy

Craco is a striking and somewhat ghostly village. In 1963, a landslide forced its inhabitants to relocate to the new municipality of Craco Peschiera, and Craco has since become a popular tourist destination for visitors from around the world. The old town of Craco still retains houses clinging to the rock, particularly the Norman tower that rises above the highest part of the village. The cluster of houses surrounding the square tower dates to the period between 1154 and 1168 during the reign of Frederick II. In 1276, the ghost town of Craco also hosted a university. Later, the town was attacked several times by brigands during the Napoleonic era, in 1807 and again in 1861. The recent history of Craco is marked by the disastrous landslide of 1963: the collapse of houses and buildings was gradual, forcing residents to abandon their homes by 1974 and move downhill to Craco Peschiera. This was not the first natural disaster affecting the village; a severe earthquake in 1688, centered on Craco-Pisticci, also caused latent landslides on this already unstable terrain. The project aims to build a new cultural attraction that enhances visits to Old Craco [31] while actively involving young people. It is part of the Heritage Smart Lab initiative, supported by the Basilicata Creativa cluster and led by the CNR of Basilicata, in collaboration with the University of Basilicata and ENEA. The project is co-funded by the Basilicata Region under the PO FESR 2014–2020 program for Research, Innovation, and Technological Development. Heritage Smart Lab seeks to develop prototypes, procedures, and operational protocols in knowledge, preservation, enhancement, and enjoyment of cultural heritage through the Smart Lab model—real development hubs involving institutions, companies, and active citizens. Among over twenty labs planned across Basilicata, Craco will play a primary role in cultural site accessibility. The "Craco Accessible" Smart Lab will implement AI-based technologies to improve socio-cultural, physical, and economic accessibility to heritage sites.

Geographical Data

Location: Craco

Country: Italy

Region: Basilicata

Language: Italian

Time zone: Rome (CET)

Altitude: 391 m above sea level

Population: 599.

Project Details and Objectives

Project funding: Co-funded project

Number of villages involved: 1

Project period: 2016–present.

The Heritage Smart Lab project aims to create a new cultural attraction in Old Craco while encouraging active youth participation. Led by the CNR of Basilicata, with support from Basilicata Creativa, the University of Basilicata, and ENEA, and funded by the Basilicata Region under PO FESR 2014–2020, the project develops prototypes, procedures, and operational protocols for knowledge, conservation, enhancement, and cultural heritage enjoyment via Smart Labs. In particular, "Craco Accessible" will implement AI-based technologies to enhance socio-cultural, physical, and economic accessibility to cultural sites, playing a central role in the project, which will activate over twenty labs in Basilicata in the coming months.

Key Themes: Accessibility, Historic and architectural heritage recovery, Structural adaptation, Cultural enjoyment and enhancement, Youth entrepreneurship.

2.3.16 Erice—Italy

Erice is a charming medieval village perched atop the eponymous mountain, located on the western coast of Sicily, under the province of Trapani, 13 km from the provincial capital. The village is characterized by a mixture of architectural styles—medieval, Arab, Baroque, and Liberty—reflecting its ancient and complex history. Within Erice, visitors can find watchtowers, monasteries, gates, a mysterious castle, and more than sixty churches, all enclosed by cyclopean walls. The village has 512 inhabitants, but its streets come alive during the summer months as tourists arrive seeking history, good food, and breathtaking beauty. In 1241, the territory of Erice was much larger than it is today, known as Agro Ericino, which included Valderice, part of Castellammare del Golfo, San Vito Lo Capo, Custonaci, and Buseto

Palizzolo. By 1846, the Agro Ericino extended over 40,000 ha with a coastline of more than 26 miles. At the end of the nineteenth century, much of the territory was assigned to Castellammare del Golfo. The village's streets are narrow and winding, opening onto panoramic terraces with stunning views of the Mediterranean Sea. Homes feature beautiful internal courtyards, offering privacy and creating a serene, almost surreal atmosphere. The main goal of the project is the creation of an online information and service system for citizens and businesses, active 24/7, allowing direct interaction with the Municipality from home. This is achieved through the new portal "Servizi ONLINE," [32] a multifunctional telematic desk that enables users to request and print personal certificates, check their tax status with a click, pay for individual-demand services online, and submit geolocated reports with photos. In short, it provides a unified access point for all municipal services and territorial information, accessible even via smartphone through the innovative application "Municipium." The Erice Smart Evo project (Action 2.2.1) aims to enhance and expand current software tools to promote interaction between the Municipality and potential users. It covers areas of services currently lacking digital access channels, directly used by citizens or their delegates.

Geographical Data

Location: Erice

Country: Italy

Region: Sicily

Language: Italian

Time zone: Rome (CET)

Altitude: 751 m above sea level

Population: 500.

Project Details and Objectives

Number of villages involved: 1

Project period: 2017–2019.

Objectives

- Request and print personal certificates;
- Consult tax status;
- Pay for individual-demand services online;
- Submit geolocated reports;
- Provide a unified access point for municipal services.

Key Themes: Enhancing village vitality, Digitalization, Tourist enjoyment.

2.3.17 Laurenzana—Italy

Laurenzana is a mountain resort town, beautifully and panoramically situated on the Camastra Valley at 850 m above sea level. It is a small jewel of art, culture, and nature, nestled among the forested mountains of the Lucanian Apennines, which reach up to 1,456 m with Mount Caperrino. The ancient medieval core is dominated by the Castle and the Mother Church, which rise on two rocky spurs. Much of the municipal territory is part of the "Appennino Lucano Val d'Agri Lagonegrese" National Park; the crown jewel of the park is the Laurenzana Fir Grove. The historic center, at the foot of the Castle and Mother Church cliffs, features a suggestive layout with narrow streets called "strèttele," stone houses, noble palaces, underpass arches of houses, and votive shrines. Largo Fiera and the nearby alleys, at the beginning of the twentieth century, formed a "village-factory" with warehouses and workshops revolving around the core of the Old De Rosa Silk Mill. Today, it stands as a testimony to industrial archaeology, with machinery, equipment, and tools preserved as they were at the peak of production, when they transformed local wool into yarns and textiles.

The "one-euro houses" initiative [33] has gained renewed attention with the deed announced by the municipality of Laurenzana in Basilicata. Unlike other participating municipalities, Laurenzana does not require a security deposit at the contract signing to support and encourage buyers, making it easier to invest without strict bureaucratic requirements. The first house was acquired by the owner of a technical studio in Casalvieri, in the province of Frosinone, who plans to transform it into a hotel to allow everyone to enjoy the beauty of the area. The "One-Euro Houses" project is aimed at anyone—Italian or foreign citizens, businesses, and associations—with the goal of recovering degraded and abandoned urban spaces, revitalizing the town. The properties offered at the symbolic price of one euro are mostly dilapidated and require renovations costing at least 80,000–90,000 euros.

Geographical Data

Location: Laurenzana

Country: Italy

Region: Basilicata

Language: Italian

Time zone: Rome (CET)

Altitude: 850 m above sea level

Population: 1,565.

Project Details and Objectives

Number of villages involved: 1

Project period: 2023.

Objectives

- Urban regeneration;
- Revitalization of the town;
- Encouragement of buyers;
- Promotion of tourism;
- Economic and social sustainability.

Key Themes: Rehabilitation of abandoned buildings.

2.3.18 Santa Fiora—Italy

Santa Fiora is a medieval village on Monte Amiata in the province of Grosseto. It has been awarded the Orange Flag by the Touring Club and is listed among the "Most Beautiful Villages in Italy." Today, however, its fame is mainly linked to remote working: it was the first town to present itself as a smart village during the pandemic, attracting residents from all over the world. This marks only the beginning of a revival project that includes opportunities for businesses and startups. With just 2,500 inhabitants perched on the slopes of Monte Amiata, Santa Fiora gained international attention (with mentions on CNN and Forbes) for offering itself as a smart working village for those seeking a work refuge during the pandemic. The success [34] is due in part to fiber-optic connectivity bridging the technological gap between rural and urban areas, and in part to the intuition of Mayor Federico Balocchi, who saw in the global health emergency an opportunity for social and economic renewal that continues today. Nearly thirty families—many from northern Italy and abroad—participated in temporary residencies over 18 months, rotating in the village's vacant homes, with some eventually relocating permanently. Santa Fiora became the first, and possibly the only, municipality in the world to offer not only individual homes but the entire town to newcomers. To encourage new temporary residents, the municipality provides financial incentives, including rental support, and a simple, clear website with all necessary information. The local population, initially skeptical, quickly adapted to the newcomers and the well-being brought by this new course. The real estate market, both in the historic center and in the surrounding hamlets, experienced a sudden revival. Previously unused second homes were restored, and local micro-enterprises began to thrive again.

Geographical Data

Location: Santa Fiora

Country: Italy

Region: Tuscany

Language: Italian

Time zone: Rome (CET)

Altitude: 687 m above sea level

Population: 2,484.

Project Details and Objectives

Number of villages involved: 1

Project period: 2 years

Project budget: €1,600,000.

Objectives

- Attract new residents;
- Promote permanent settlement;
- Provide economic incentives;
- Offer an informative website;
- Revitalize the real estate market;
- Support local micro-enterprises.

Key Themes: Smart working.

2.4 Critical Synthesis of European Smart Villages

The analysis of European villages highlights strong heterogeneity in size, development strategies, and adopted technologies, while sharing some common traits linked to heritage valorization, community cohesion, and sustainability.

1. **Small villages and demographic challenges**: Many European villages, such as Ferraria de São João (Portugal), Tomaszyn (Poland), Civita di Bagnoregio, Craco, Erice, Laurenzana, and Santa Fiora (Italy), have small or declining populations, often with high aging indices. In these contexts, the smart village projects focus on attracting new residents, promoting sustainable tourism, and encouraging youth retention. Urban regeneration, public space redevelopment, and cultural heritage valorization are strategic tools for social and economic revitalization.

2. **Culture- and heritage-based approaches**: Villages such as Phini (Cyprus), Teruel and Laurenzana (Spain and Italy) emphasize strengthening cultural identity and traditions through educational initiatives, festivals, workshops, and restoration courses. Local craftsmanship, typical products, and historical architecture are leveraged not only for tourism but also to stimulate local economy and community cohesion.
3. **Technological innovation and digitalization**: Some villages adopt digital technologies to improve public services, communication, and economic opportunities. Remetea (Romania) and Erice (Italy) provide examples of e-governance platforms and mobile apps for municipal services, while Raudanmaa (Finland) and Vuollerim (Sweden) integrate digital technologies to support social and economic innovation. Tomaszyn (Poland) implements an integrated approach with smart farming and closed ecosystems, combining technology and environmental sustainability.
4. **Social innovation and community participation**: Most villages emphasize active citizen engagement: Ferraria de São João promotes volunteering and participatory tourism; Vuollerim and Phini develop cooperative models and participatory forums; Stanz (Austria) and Santa Fiora involve citizens in service management and urban regeneration. Social inclusion is often the core of the smart strategy, complementing technological interventions.
5. **Environmental sustainability and resource management**: Many villages focus on sustainable resource management and energy efficiency. Zaragoza promotes shared photovoltaic energy, Raudanmaa integrates sustainable water networks and fiber optics, and Santa Fiora uses digital connectivity to support smart working, reducing territorial impact. Ferraria de São João emphasizes sustainability in agriculture, tourism, and forest management.
6. **Diversified strategies**: Larger or better-funded villages (Zaragoza, Remetea, Civita di Bagnoregio) adopt complex strategies combining digitalization, cultural tourism, education, and social inclusion, whereas smaller villages (Tomaszyn, Laurenzana, Santa Fiora) often focus on a few key projects such as urban regeneration, smart working, or sustainable agriculture.

European cases in Table 2.2 demonstrate that there is no single model for a smart village: strategies must adapt to local characteristics, population size, and available resources. However, three recurring levers are crucial for successful projects: heritage and cultural valorization, social and participatory innovation, and intelligent integration of technology. The most effective experiences combine these elements in an integrated way, aiming for economic, environmental, and social sustainability.

Table 2.2 European smart villages

EU Village	Country	Population	Altitude (m)	Project period	Main objectives	Key themes	Distinctive features
Ferraria de São João	Portugal	43	Mountain peak	2020–2029	Strengthen rural identity, promote tourism and social cohesion, heritage valorization	Culture, sustainable tourism, social cohesion	High community engagement; dynamic in agriculture and tourism
Phini	Cyprus	391	1,000	2019–present	Promote culture and traditions, youth participation, mountain area policies	Culture, tourism, economic development, education	Annual thematic conferences; strong school and community engagement
Raudanmaa	Finland	510	121	2017–2019	Innovation and services to improve daily life, local economy	Social innovation, quality of life, culture	Smart Town Hall; digital cooperative; teleworking
Remetea	Romania	6,261	750	Until 2021	Counter youth migration, digitalize services, online education	E-governance, digital education, local development	RemeteApp; digital and educational platforms
Zaragoza	Spain	682,513	208	Until 2021	Sustainable energy, social inclusion, community participation	Community energy	Shared photovoltaic energy; fight against energy poverty
Stanz	Austria	441	1,040	Until 2021	Self-sufficient energy, sustainable tourism, civic participation	Urban regeneration, sustainable mobility	"Mitfahr-Bankerl"; multifunctional urban center

(continued)

Table 2.2 (continued)

EU Village	Country	Population	Altitude (m)	Project period	Main objectives	Key themes	Distinctive features
Tomaszyn	Poland	18	135	21 months	Organic farming, closed-loop ecosystem, economic resilience	Smart farming, environmental sustainability	Ostoja Natury 3.0; closed-loop system
Vuollerim	Sweden	732	400–500	20+ years	Local economic development, cooperation, youth opportunities	Community cooperative, quality of life, rural innovation	Focus on cultural tourism and Saami experiences
Teruel	Spain	35,675	915	Present	Heritage restoration, cultural promotion	Restoration, artisan training	Fundación Santa María de Albarracín; educational workshops
Civita di Bagnoregio	Italy	3,303	200	2011–present	Heritage protection, research, education	Landscape conservation, sustainable tourism	Agreement with IUAV; Etruscan heritage valorization
Craco	Italy	599	391	2016–present	Cultural tourism, youth engagement	Accessibility, heritage, innovation	Craco Accessible Smart Lab; AI for cultural sites
Erice	Italy	500	751	2017–2019	Digitalization of services, citizen–municipality interaction	E-governance, tourism, vitality	Online services and Municipium app

(continued)

Table 2.2 (continued)

EU Village	Country	Population	Altitude (m)	Project period	Main objectives	Key themes	Distinctive features
Laurenzana	Italy	1,565	850	2023	Urban regeneration, tourism, attracting buyers	Building restoration, sustainable tourism	"1-euro houses"; historical center valorization
Santa Fiora	Italy	2,484	687	2 years	Attract new residents, smart working, micro-enterprises	Smart working, property regeneration	First Italian smart working village; economic incentives

References

1. Zhang, X., & Zhang, Z. (2020). How do smart villages become a way to achieve sustainable development in rural areas? Smart village planning and practices in China. *Sustainability, 12*(24), 10510. https://doi.org/10.3390/su122410510
2. The Central Committee of the Communist Party of China and the State Council issued the "Strategic Plan for Rural Revitalization (2018–2022)". Retrieved September 26, 2020, from http://www.moa.gov.cn/ztzl/xczx/xczxzlgh/201811/t20181129_6163953.htm
3. The General Office of the Central Committee of the Communist Party of China and the General Office of the State Council issued the "Digital Rural Development Strategy Outline". Retrieved May 16, 2019, from http://www.gov.cn/zhengce/2019-05/16/content_5392269.htm
4. Aziiza, A., & Susanto, T. (2020). The smart village model for rural area (case study: Banyuwangi Regency). *IOP Conference Series: Materials Science and Engineering, 722*, 012011. https://doi.org/10.1088/1757-899X/722/1/012011
5. Anwar, A., Pamungkas, T. S., Wicaksono, G., & Suharsono, A. (2019). Innovation of local village budgeting management: "The Case Study of Banyuwangi Regency Electronic Village Budgeting". Regional Dynamic: Journal of Policy and Business Science, *1*(1).
6. Khusaini, M. (2015). A shift-share analysis on regional competitiveness-a case of Banyuwangi district, East Java, Indonesia. *Procedia-Social and Behavioral Sciences, 211*, 738–744. https://doi.org/10.1016/j.sbspro.2015.11.097
7. World Bank Group. (2021). *Niger: Smart villages for rural growth and digital inclusion—Procurement plan (English)* (Report No. P167543). Author. https://documents.worldbank.org/curated/en/933441623264623448
8. International Telecommunication Union (ITU). (2020). *Building smart villages: A blueprint as piloted in Niger*. ITU. https://www.itu.int/
9. Talbot, D., & Paz-Canales, M. (2017). *Smart grid paybacks: The Chattanooga example*. The Municipal Fiber Project, Berkman Klein Center for Internet & Society Research Publication.
10. Martinez Salgado, J. D., Borrayo, A. S., Claudia, L., Castellanos, A., & Martinez Baron, D. (2025). *The climate-smart villages (CSV) of Olopa as a local innovation platform for scaling ecosystem-based adaptation in Guatemala* (p. 9).
11. Pokharel, R., Mishra, A. K., & Aithal, P. S. (2021). Practicability assessment of smart village project: A case of Sandakpur rural municipality, Ilam Nepal. *International Journal of Management, Technology, and Social Sciences (IJMTS), 6*(2), 265–281. https://doi.org/10.47992/IJMTS.2581.6012.0170
12. Yoon, C., Lim, D., & Park, C. (2020). Factors affecting adoption of smart farms: The case of Korea. *Computers in Human Behavior, 108*, 106309. https://doi.org/10.1016/j.chb.2020.106309
13. World Bank. (2021). *Smart villages in Azerbaijan: A framework for analysis and roadmap*. © World Bank. http://hdl.handle.net/10986/35468
14. Hashimoto, A., Telfer, D., & Telfer, S. (2020). Life beyond growth? Rural depopulation becoming the attraction in Nagoro, Japan's scarecrow village. *Journal of Heritage Tourism, 16*, 1–20. https://doi.org/10.1080/1743873X.2020.1807556
15. Erin, P. (2020). Community in rural self-government of the Russian village before and after 1917 (based on materials from the Tambov Governorate). *Tambov University Review. Series: Humanities*, 175–180. https://doi.org/10.20310/1810-0201-2020-25-189-175-180
16. Zavratnik, V., Kos, A., & Stojmenova Duh, E. (2018). *Smart villages: Comprehensive review of initiatives and practices*. MDPI.
17. Smart Rural 21. (2020, December). *Smart village strategy of Alsunga (Latvia)*. Preparatory Action on Smart Rural Areas in the 21st Century. https://www.smartrural21.eu/wp-content/uploads/Alsunga_Smart-Village-Strategy.pdf
18. 2nd Preparatory Action—Smart Rural Areas in the 21st Century. (2023, November 27–28). *Report on 2nd European smart villages pilot observatory meeting, Prague, Czech Republic*. Funded by the European Union (Contract No: AGRI).

19. Smart Rural 21. (2020, June). *Smart village strategy of Kythera (Greece).* Preparatory Action on Smart Rural Areas in the 21st Century.
20. Smart Rural 27. (2021). *Smart villages in the CAP strategic plan: France* [Factsheet]. European Commission.
21. Municipality of Penela. (2020, December). *Smart village strategy of Penela—Ferraria de São João and Cumeeira, Portugal.* Preparatory Action for Smart Rural Areas in the 21st Century. European Commission.
22. Smart Rural Areas. (2024). *Smart villages in Cyprus: Planning and implementation under the Smart Rural 27 project.*
23. European Union. (2023). *Raudanmaa—Finland: Smart village strategy.* Smart Rural 21.
24. E40 Consortium. (2020). *Smart village strategy of Remetea.* Prepared in the framework of the 'Preparatory Action on Smart Rural Areas in the 21st Century' project funded by the European Commission.
25. García Fernández, C., & Peek, D. (2023). Connecting the smart village: A switch towards smart and sustainable rural-urban linkages in Spain. *Land, 12*(4), 822. https://doi.org/10.3390/land12040822
26. Smart Rural Areas. (2021). *Stanz im Mürztal: Smart village strategy and energy community.* Smart Rural Areas.
27. Smart Rural Areas. (2021). *Tomaszyn village strategy: Ostoja Natury and regenerative agriculture.* Smart Rural Areas.
28. Smart Rural 27. (2021). *Vuollerim village strategy.* Smart Rural 27.
29. Rodrigo-Ilarri, J., Rodrigo-Clavero, M.-E., Romero-Hernández, C. P., Bernad-Esteban, P., & Benito Ruiz, E. (2025). Comprehensive proposal for the rehabilitation of the Acequia del Diablo (Teruel, Spain): Revitalizing irrigation and cultural heritage. *Sustainability, 17*(10), 4519. https://doi.org/10.3390/su17104519
30. Associazione Civita. (2011, Aprile 20). *Civita di Bagnoregio non muore più.* Archeomatica. https://www.civita.it
31. Basilicata Creativa ETS. (2025, Ottobre 1). *"Craco accessibile": il borgo fantasma diventa interattivo grazie a un nuovo applicativo digitale.* Basilicata Creativa. https://www.basilicatacreativa.it
32. Città di Erice. (2019, Gennaio 14). *Erice Smart: il 22 gennaio la presentazione del nuovo Portale "Servizi online".* Regione Siciliana. https://www.comune.erice.tp.it
33. Gualtieri, P., & Dallagiovanna, G. (2021, April 7). *Laurenzana, in Basilicata, dove comprare case a 1-euro senza deposito.* Immobiliare.it News. https://www.immobiliare.it/news/laurenzana-case-1-euro-senza-deposito
34. Comune di Santa Fiora. (2025, April 2). *Santa Fiora smart village.* Regione Toscana. https://www.comune.santafiora.gr.it/attuazione-misure-pnrr/santa-fiora-smart-village

Chapter 3
Strategies for the Rehabilitation, Safeguarding, and Enhancement of Rural Areas Through a Smart Approach

Abstract This chapter presents a systematic strategy for the rehabilitation, safeguarding, and enhancement of rural areas through a smart approach that combines technological and digital innovation with green and circular economy solutions. Social innovation and active community participation are central, integrating governance, heritage preservation, technology, and digitalisation. Guidelines, best practices, management protocols, and business plans are provided to support implementation, monitoring, and long-term sustainability. The chapter offers a practical roadmap for creating Smart Villages that merge innovation, community engagement, and environmental responsibility.

In the previous chapter, *Smart Villages* and policies for inner areas were examined, from national structural initiatives to micro-scale interventions, highlighting the tools supporting small communities and intervention strategies. Numerous studies in the area in recent years have attempted to explore the dynamics and processes related to small rural communities. These processes, as emphasized by Scoones, develop in two areas: that of policies and that of subsistence, where political choices can favour some groups and disadvantage others [1]. He also highlights the need to investigate the "black box" of institutions and public policies. Fraser, on the other hand, emphasizes the importance of obtaining recognition and promoting participation to achieve citizens' emancipation, initiating a process of citizen empowerment [2–4]. Active participation in community life and the ability to claim political rights are essential for individual well-being [5, 6]. In this context, it is crucial to understand the role of local communities in recreating connections and projects. As highlighted by Chambers, communities are not only bearers of interests but co-creators of development, challenging the *top-down* approach that prevailed during the Fordist period [5, 7].

Following this line, Scoones draws on Chambers to outline a paradigm shift in rural development. He asks: "*Whose knowledge truly matters? Which livelihood strategies are taken as models when formulating intervention policies?*" [1, 5]. It is with this perspective that the experimentation described in this chapter aims to delve into the micro-processes, working from within the communities to provide concrete

L. Lombardo, *Smart Villages*,
SpringerBriefs in Applied Sciences and Technology,
https://doi.org/10.1007/978-3-032-16031-7_3

responses to *governance*, and thus to inner areas, using innovative and sustainable approaches.

In recent years, the SNAI has promoted, albeit to a minimal extent, the protagonism of local communities, challenging the closure of some local realities that hinder innovation [8]. The strategy assigns the local community the task of transforming its specificities into projects, enhancing the work of innovating actors who, although sometimes operating in isolation, are connected to supra-territorial networks. An analysis of the strategic documents approved until 2020 highlights how the word "Community" is central, although it takes on a romantic and idealized character, as emphasized by DeFilippis [9]. It is more about a new community than a pre-existing form, for which it is necessary to build "voluntary and open" processes capable of creating new identities [10]. The definition of community by McMillan and Chavis [11] emphasizes the importance of social bonds between different individuals engaged in common actions in local contexts. However, this definition does not explore the potential of communities in development processes. The literature presents two opposing views: one that celebrates the community as the centre of social initiatives, and the other that criticizes its political effectiveness [9]. DeFilippis proposes a more nuanced understanding, seeing the community as a product of its broader contexts but also as an arena for social change. Thus, the community is seen as a political space where significant change can occur. Olivetti defines the community as the site of political elaboration and social mediation, an independent entity that does not require political parties [12]. This approach, while radical, can be reinterpreted as a pragmatic response to contemporary challenges. Communities are not static entities but can evolve and respond to needs, as highlighted by Cocco [13], and creating opportunities in the territory. Experiences of community cooperatives and shared management of common goods, for example, often initiated by small groups, demonstrate how structured responses to local needs can be created, generating new perspectives in terms of relationships, job opportunities, but above all, social impact.

In Sardinia, community cooperatives were born in response to local challenges, with the support of municipal administrations, as in the cases of *Fluminimaggiore* and *Ollolai* [14]; activities that are still not widely known or explored in Sicily. However, many policies continue to be imposed from the top down, with limited citizen participation, unlike in Europe, where the sense of community and activism for the regeneration of places are much more deeply rooted and create connections and projects. Community cooperatives are just one of the many examples that can be undertaken at the local level, one of the many possible strategies for the rehabilitation and enhancement of small rural contexts.

Today, sustainability and the enhancement of smaller centres are priority objectives that a small community must approach. At the same time, it is also close to these goals because it operates in well-preserved and already inherently sustainable contexts. However, to face this challenge, it is necessary to redefine the "theory of value," adopting an approach that integrates the quality of the context, sustainable design, and technical, technological, and practical feasibility, so that economic, environmental, and social aspects are central to the territorial redevelopment and in synergy with the sustainable objectives of the 2030 Agenda.

Smart Villages represent a transition towards rural communities that adopt digital technologies and sustainable solutions, improving quality of life and promoting economic, social, and environmental development. They aim to create synergies between cultural and natural heritage and innovation, enhancing the community through participatory *governance* focused on defining strategies and objectives that are balanced with the system in which they operate. They are based on key principles such as inclusion, sustainability, and innovation that respects tradition while looking to the future. These principles translate into practical actions such as the expansion of broadband, the digitisation of public services, and the use of advanced technologies like artificial intelligence and the Internet of Things (*IoT*) to optimise resource management and improve quality of life. *Green* and circular economy, local production, rehabilitation, reuse, recycling, accessibility (both physical and digital), and *inclusive design* for resilient territorial development and intelligent, resilient communities [15].

In addition, energy efficiency, seismic adaptation, and the analysis of the technical characteristics of historic buildings are fundamental themes for imagining and rethinking the basic infrastructure that characterises these small rural contexts. These priorities are highlighted, as an example, in the proposal of the Atlas and Guidelines for the Municipality of *Polizzi Generosa* and, by analogy or contrast, can be applied to other rural contexts. Ensuring that the building heritage meets high living standards is one of the pillars of *Smart Villages*. Advanced technologies enable the reduction of environmental impacts from buildings and improve citizens' well-being by monitoring energy consumption in real time and contributing to the reduction of greenhouse gas emissions [16]. However, all these activities prioritise tourism development, which remains one of the major pillars capable of generating economic income. The risk, however, is that it should not become the sole driver of the local economy but rather be integrated into a broader context of enhancing the territory's resources; this is to avoid phenomena of *touristification* that could erode the essence of well-preserved villages and contexts, such as the small inner towns.

Looking at the training process in the management of *Smart Villages*, it is therefore appropriate to follow precise steps and conduct an in-depth analysis of the territory, aimed at identifying the strengths, weaknesses, opportunities, and threats within the area. SWOT analysis, questionnaires, and surveys allow for the precise identification of specific needs and opportunities for improvement in the local context. This preliminary phase provides a solid foundation for identifying priorities, needs, and resources present or necessary to develop a successful project. Alongside this, it is essential to promote interdisciplinary collaboration, involving professionals from various sectors, such as urban planners, engineers, economists, and sociologists, as well as universities and research institutions. A holistic approach integrates diverse expertise, ensuring that solutions are not only sustainable but also feasible in the long term.

Another crucial element in managing *Smart Villages* is the building of strong public–private partnerships. Collaboration between the public and private sectors allows access to specific expertise and strategic investments, contributing to the creation of a shared data system aimed at achieving common goals. This approach strengthens the capacity to implement innovative solutions and ensure the economic sustainability of the project. *Guidelines, best practices, business plans*, and *management protocols*, alongside regular monitoring, are other tools that a *Smart Village* must consider for development. In this context, adherence to the principles of projects such as the National Rehabilitation and Resilience Plan (PNRR) is a priority. However, as with all European funds, it is essential to keep competencies up to date to respond to new dynamics and emerging needs, both at the community and global levels, going beyond the 2030 Agenda and the Green Deal, while considering the possibility of accessing as many funds and projects as possible to expand the range of potential solutions for the territory.

In well-preserved contexts (Association of Virtuous Municipalities), the enhancement and protection by institutions such as the Superintendence, extended to both historic centres and landscapes, serve as guiding principles for rural contexts within UNESCO Global Geoparks or territories of great natural value, not only for the area but, above all, for humanity. These principles and processes, as well as the themes to be developed, ensure that every innovation aligns with protocols and regulations for the safeguarding of cultural and natural heritage, with a constant commitment towards sustainable solutions.

This varied growth path integrates advanced technologies with respect for tradition, projecting the territory towards a sustainable, inclusive future rich in new opportunities; it is not about being tied to the past, but rather about seeking to transpose and project it into the future. The main objective that goes beyond the project of a *Smart Village* in a single village remains, however, the creation of a network of *smart villages*, an interconnected system that promotes sustainable development at the local level and, gradually, extends from the single municipality to the broader areas (Bobbio, 2004). At the heart of the future network vision is the creation of an exchange platform, where *smart villages* can share experiences, ideas, and solutions to replicate in other contexts, an ideal platform capable of providing the tools necessary for the protection and enhancement of territories. It is not about creating a Smart Village but about promoting and contributing to the formation of united, collaborative villages that drive shared initiatives. Such a platform not only strengthens collaboration but also creates synergies between communities, fostering mutual learning and growth according to the principle that "together we are stronger." Through this network, each village benefits from the *best practices* of others, but above all, each village, with its peculiarities, can contribute to filling the gaps of others, mutual aid and *co-design*, accelerating innovation processes and adaptation to local challenges.

A crucial element of this approach is the enhancement of the local ecosystem: each village is therefore encouraged to develop and support initiatives in line with the specific needs of its territory, its vocation, creating an environment in which innovation thrives and integrates organically into the social and cultural fabric without disrupting it. This approach ensures that technologies are implemented in a relevant manner, responding to local challenges, and that there are no forced or inappropriate solutions.

The proposed interventions are conceived as scalable models, adaptable not only to Sicilian realities but also to national and European contexts, respecting key principles and objectives in line with European Union policies. The proposed methodology aims at the widespread dissemination of the process of creating *Smart Villages*.

The discussion concludes by providing concrete support to the reality analysed, as well as to other contexts by analogy or contrast. Practical tools for the technological and compatible rehabilitation of the territory, as well as business models tailored for small municipalities, will be promoted to ensure conscious project planning. To conclude this journey, guidelines, *best practices*, *business plans*, and *management protocols* will ensure the sustainable development of territories.

All these tools are meticulously described in this chapter with the goal of outlining a shared and pragmatic path for the development of *Smart Villages*, following the parameters of protection and conservation, rehabilitation of existing heritage, and compliance with regulations. However, they must adopt a virtuous path and achieve goals as outlined in the provided diagram. In detail, in these months, a map has also been developed, inspired by representations of *Smart Villages* around the world, particularly in Sweden, and their representation method, identifying strategies and territorial opportunities that can help various contexts maintain a status of "relevance" by promoting territorial characteristics [17] and a shared path where technologies converge in the sustainability process. This map has been tailored to a Sicilian context (Madonie district), identifying different smart strategies to be applied for each municipality in the rural context. The metro map (Fig. 3.1) developed is inspired by the traditional metro schemes found worldwide. Specifically, it unfolds in three distinct stages: firstly, it identifies the rural centres, mapping them accurately within the territory. Secondly, it connects these centres through a network of smart and technological paths, all oriented towards the central track of sustainability. Finally, for each theme, the map proposes a specific route to follow for achieving smart objectives. The points of intersection or intermediate stops represent the necessary stages to meet the defined parameters for *Smart Villages*; this is an initial approach to outline directions and paths to undertake and pursue.

3.1 The Stakeholders of a Smart Village

The As outlined in the previous paragraph, the path to becoming a *Smart Village* is not simple and primarily requires the establishment of a system capable of supporting all subsequent steps. The concept of a *Smart Village* revolves around the integration of various actors who, with their complementary contributions, enable the territory to grow sustainably, enhancing existing heritage and responding to local needs through innovative and shared solutions. Over the years, numerous opportunities have been offered by research worldwide on the perception and functioning of *Smart Villages* [18]. However, this research aims to focus on the importance of built heritage rehabilitation and the awareness that an integrated system of actors and solutions can serve as a model for creating a new type of Smart Village, based on the understanding

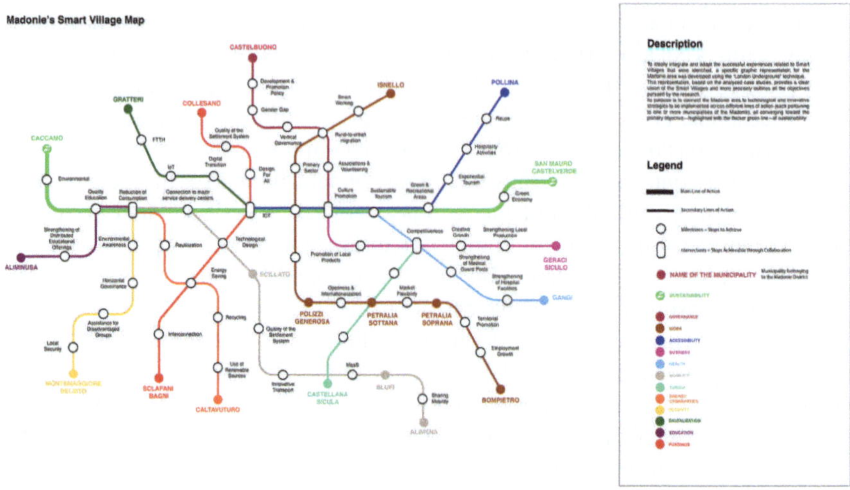

Fig. 3.1 Madonie's smart village maps. © Image by author, 2024

that the integration of tradition and innovation can coexist and drive its perfect and balanced functioning, resulting in sustainable outcomes in both the short and long term.

To build a solid structure, it is necessary to have an organization that is equally effective and integrated. Therefore, it is useful to consider two fundamental spheres of intervention: the Social and Participatory Component and the Technological and Heritage Component. These two spheres reflect a dual approach, considering, on one hand, human capital and the active participation of the entire community, and, on the other, the use of advanced technologies, digitalisation, and attention to the rehabilitation, enhancement, and protection of heritage.

Social and Participatory Component

The social and participatory component is the human and relational core of the Smart Village, crucial for fostering a sense of belonging and cooperation within the local community [19]. It is divided into two essential subcategories:

- community: focuses on the active involvement of citizens in decision-making processes and in the daily life of the village/town. In this perspective, the community is not just a group of individuals, but a cohesive and collaborative social network [20], where participation and cooperation are encouraged to strengthen the identity bond and generate a sense of collective belonging. In a *Smart Village* model, the community can and should be engaged through participatory tools or processes, or by forming cooperatives of any kind (energy or community-based), which, in addition to stimulating the local economy, promote a sense of ownership over shared projects and initiatives. These tools are essential for creating local employment, enhancing resources, and ensuring long-term sustainability;

- governance: represents the coordination, administration, and management element within the Smart Village. It includes both local institutions and administrations, with its primary role being to facilitate transparent and constructive interaction with the community [21]. Good *governance* is characterized by the ability to actively listen to the needs of the population and integrate local demands into public policies. Transparency, inclusivity, and accountability are essential principles that guide *governance* actions, which should aim to build trust, strengthen social cohesion, and support economic and social activities promoted by the community itself. Essentially, *governance* represents the management of resources and strategic actions that make the development objectives of a *Smart Village* a reality. Moreover, *governance* has the role of identifying the needs encountered and translating them into projects and programmatic actions, both through European projects and any available tools.

Technological and Heritage Component: focuses more on outcomes, the modernization of services, and the protection of architectural, historical, and cultural heritage, combining tradition and innovation to ensure that progress is in perfect harmony with local identity and compatible with the built heritage. This component is structured in the following areas:

- technology and digitalisation: The use of technology is a fundamental aspect for improving the efficiency and accessibility of public services and for facilitating communication and interaction between the community and *governance.* In the context of a Smart Village, digitalisation helps to reduce the gap between citizens and administration, allowing for more flexible and transparent management of resources. The integration of digital solutions, such as online platforms for accessing services or monitoring systems for managing environmental resources, contributes to optimizing resource use and improving quality of life. Similarly, technologies applied to the building heritage make the current building stock more relevant and aligned with contemporary needs. This technological approach promotes "*Smart Governance*," which leverages innovation to respond efficiently and sustainably to the needs of the local population.
- rehabilitation and enhancement of material and immaterial heritage: Another crucial aspect of the *Smart Village* is the rehabilitation and enhancement of heritage, which includes not only the protection of architecture and the conservation of local identity characteristics but also nature and intangible heritage [22]. The preservation and rehabilitation of architectural, historical, cultural, and environmental heritage help strengthen local identity, making the traditions and values of the territory an integral part of contemporary development [23]. The enhancement of intangible heritage, such as customs, traditions, and traditional practices, is a key element in ensuring that technological progress is accompanied by cultural continuity [24]. This perspective enhances the authenticity of the territory and ensures that modernization does not compromise but rather enriches the local peculiarities and traditions.

The interconnection between these two components is not only desirable but necessary for the balanced and sustainable development of a Smart Village. The active participation of the community [25], supported by attentive and transparent *governance*, represents a solid foundation for the implementation of technological innovations which, in turn, contribute to improving the efficiency and quality of services, making life easier for residents. Furthermore, technology is functional to the protection of heritage: through monitoring and digital management systems, it is possible to protect and enhance architectural, environmental, and cultural resources [26], enabling them to be understood and shared, while ensuring that cultural, economic, and social growth is sustainable and respectful of local identity. A *Smart Village* must, therefore, be understood as an integrated ecosystem, where technological innovation harmoniously blends with the enhancement of human capital and local heritage, creating a development model that is inclusive, resilient, and able to respond effectively to the contemporary challenges faced by rural and inner areas. Within a Smart Village, the community plays a crucial role in the enhancement and protection of heritage, being the main party responsible for its conservation and promotion. Raising awareness and educating the local population are fundamental to ensuring informed and lasting engagement [27]. Through cultural initiatives, training activities, and workshops, the community is and must be informed about the importance of its heritage, promoting widespread awareness that encourages respect for and protection of local resources [28]. At the same time, it is essential to encourage active participation from citizens in the design and implementation phases of initiatives [29]. *Co-design* mechanisms, as well as surveys and focus groups, allow the population to directly contribute to planning, bringing out needs, ideas, and proposals [30]. This approach generates a sense of belonging and shared responsibility, motivating the community to care for the territorial heritage. The creation of community and energy cooperatives represents another tool to strengthen this process. These cooperatives allow for autonomous and shared management of resources, contributing to local economic development and promoting a form of social economy based on the very heritage resources.

In terms of *governance*, the development of protection policies is a key step in ensuring heritage preservation [25]. Institutions, responsible for managing resources, must implement appropriate regulations, tools, and design support, encourage protection initiatives, offer financial support, and find the necessary funding to promote local enhancement projects. These policies aim to create an inclusive and sustainability-oriented regulatory framework in which heritage can be continuously protected and enhanced and receive subsidies for maintenance. Another central element is the collaboration between the public and private sectors, which enables the integration of external skills and resources in a sustainable and innovative management approach. The involvement of local actors and businesses provides new economic and technical resources, contributing to the success of enhancement projects, while constant monitoring and evaluation of initiatives allow the effectiveness of interventions to be analysed and improvements to be made. Through a transparent evaluation process, *governance* can document the impact of the actions taken, further encouraging the involvement of the community and local partners, as well as evaluating the support of universities and third-sector entities to improve the management of larger projects.

From a technological standpoint, the digitalisation of heritage represents an effective strategy to increase accessibility and promote direct and informed interaction with the public [31]. The creation of digital archives, interactive maps, and dedicated applications are just some examples that allow the community and tourists to explore and learn about local heritage in an innovative way, transforming the experience into an immersive learning opportunity. Similarly, mapping and monitoring the heritage both environmentally and structurally using advanced sensors allows the preservation of architectural, cultural, and natural assets, anticipating potential damage from wear or adverse weather conditions and taking timely action on heritage [32]. Restoration and maintenance interventions on tangible heritage must respect traditional techniques and use local materials to ensure historical and cultural consistency while minimising environmental impact and preserving the authenticity of the heritage [33]. This type of monitoring reduces the need for invasive interventions, ensuring proactive and sustainable protection. The use of digital platforms for civic participation, in turn, helps facilitate dialogue between administration and community, making it easier for citizens to report issues, propose ideas, and participate in public consultations inclusively. Finally, the enhancement of tangible and intangible heritage is the heart of any intervention in a *smart village* context.

The promotion of local traditions and skills is essential to keep the community's cultural identity alive, fostering the transmission of knowledge and practices to future generations through schools, workshops, and cultural events. These initiatives are just a few examples that strengthen the bond between the community and its territory, ensuring that local knowledge and skills can endure over time. Similarly, the creation of events and cultural routes represents an effective strategy to showcase heritage and attract conscious and sustainable tourism that respects and enhances local identity. By organising themed itineraries and dedicated events, heritage becomes not only a tourist attraction but also an opportunity to strengthen the community's sense of belonging [34]. These actions, carried out in synergy by various *stakeholders*, foster the sustainable development of the Smart Village, where the community, *governance*, technology, and heritage itself integrate to promote effective protection of local resources (Fig. 3.2). The result is a coordinated system that not only preserves the cultural and environmental identity of the territory but also generates economic and social value, building a more aware and respectful future rooted in its origins. The next section will identify, specifically, the components of a Smart Village, along with their roles and responsibilities, as well as what is identified as the method for identifying a Smart Village.

3.1.1 Smart Communities

The community represents the heart of a Smart Village, where active participation becomes essential to guide decisions in line with local needs. In this context, it

Fig. 3.2 Actors, themes, and operational indicators of smart villages. © Image by author, 2024

is crucial to educate both the involved community and the one still to be engaged, understood as an integrated system of residents, tourists, and new inhabitants, towards shared values. This education promotes the creation of shared projects, ensuring the permanence of both new and old generations in rural areas, a process that contributes to "restanza". It is important, however, to evaluate how the concept of community has changed today. One should not think of a settled community, but rather of a diversified community made up of residents, tourists, and potential new residents, including migrants, who could bring great opportunities to the area. Tools of participatory democracy generate ideas, projects, and induced tourism, and support economic management aimed at repopulating and revitalizing rural areas, not only to attract tourists [35], but also to encourage a return to origins. Local participation, in fact, strengthens territorial identity, making local policies more robust. Involving the community in public consultations, *co-design*, and shared management of services ensures inclusive *governance* and planning that responds to local needs [36]. These participatory democracy processes bring economic benefits through management models focused on community well-being and cultural heritage enhancement [37].

Participatory tools offer an effective way to engage the population in the planning and management of local resources and projects to be developed in the territory. Some concrete examples developed during the doctoral journey, in the case study of Polizzi Generosa, concern participation in the participatory democracy system and proposing solutions for the municipality, such as the implementation of green walls or the illumination of the characteristic "*tocchi*" of Polizzi, which not only offered *green* or technological solutions applied to the built heritage but also served as a reminder to raise the population's awareness and educate them about what can be done in local architectural heritage to recover and enhance it. These processes allow citizens to express their preferences and actively contribute to defining priorities for territorial development, as well as create awareness and information, as happened with the *Officine Sperimentali*, an example already developed during the SNAI period, although not widely spread as a virtuous practice to implement.

In Italy, various communities have adopted this approach, aiming to promote flexible local development processes that trigger a growth dynamic rather than implementing individual projects. The *smart community* model offers a holistic vision of territorial development that integrates cultural, architectural, and environmental heritage with digital and technological innovation [38]. This participatory context allows residents to take part in decisions concerning the environment and ongoing projects, whether under development or already completed. Through tools such as surveys and public assemblies, administrations can listen to and respond specifically to the needs expressed by the community, facilitating a participatory approach rather than a *top-down* one [39]. This method is particularly relevant in inner areas, where the aging population and geographic isolation pose significant challenges to the sustainability of services and infrastructure, requiring that projects be genuinely accessible to citizens and not imposed from above [40]. Participatory tools adopted in Amsterdam and Barcelona today involve citizens in strategic administrative decisions; members of the Cities Coalition for Digital Rights, along with Milan, launched this initiative in 2016 to promote and monitor citizens' rights.

Within this vision, citizens can become makers and/or k-makers (knowledge creators), playing a key role in the transformation process. Makers contribute their creativity to developing innovative solutions through *co-design* practices. In this way, the territory benefits from an active community that participates in the *co-design* and implementation of solutions, making local policies more effective, representative, and aligned with the needs of the area. Similarly, citizens are encouraged to identify local opportunities and understand possible steps to take independently to create new opportunities for themselves and, at the same time, create new opportunities for the territory. Involvement of citizens in the design and implementation phases of policies facilitates alignment between community expectations and sustainable development goals, contributing to inclusive and transparent *governance* [41].

Community cooperatives and energy communities, for example, represent essential tools for promoting social inclusion and strengthening social capital. Community cooperatives, often established in recent years, not only encourage the local economy but foster a sense of belonging and inclusion, mobilizing economic resources and skills for projects benefiting the community. These cooperatives, rooted in the local context, support the development of tourism, cultural, and environmental services, in line with the territorial vocation and local traditions, strengthening the link between the economy and active participation. Verbumcaudo in Polizzi Generosa is a virtuous example of resilience and opportunity in the Madonie area, turning a place once taken from the mafia into a site of production and activities for the region. They represent a bottom-up enterprise model where citizens participate in managing economic activities aimed at protecting and enhancing local heritage, promoting collective management of resources and services, with a long-term perspective that combines economic sustainability and social responsibility [42]. Finally, energy communities promote energy self-sufficiency by exploiting local resources for clean energy production [43]. This type of project not only reduces environmental impact but also creates new economic opportunities for residents, allowing the community to directly benefit from the region's resources and contributing to sustainable development [44]; thus, they become both producers and users. The effective integration of these strategies into public policies demonstrates how community participation can support high-impact social, energy, productive, and economic projects, ensuring an improvement in quality of life and adequate protection of local heritage, as highlighted by the many opportunities offered by the Atlas [45]. In Sicily, the European project SIAS2 demonstrated how community participation can inspire initiatives and projects tailored to the community, including social enterprises that create job opportunities, accessible social and cultural services, and civic participation in territorial planning, creative and training hubs, craftsmanship, and local production.

Agricultural cooperatives and local communities also emerge as solutions for agricultural, social, and economic sustainability and local food production, through resource sharing and direct sales of slow-food products, which are, in most cases, produced by locals using traditional practices. *Innovation hubs* and social factories represent collaboration spaces for developing new ideas and strengthening the local economy, not by creating ad hoc structures, but by considering the systematization of local resources to support possible activities to be launched [46]. Sustainable tourism projects such as agritourisms and hiking trails, experiential and slow tourism, enhance natural and cultural resources, bringing economic benefits to rural communities. Training programs and skill development are essential for retaining residents in the communities and contributing to their development [47–49].

Starting design processes without active community involvement in management, administrative, and decision-making mechanisms risks depriving the project of meaning for the inhabitants, who would not feel like an integral part of the initiative [50]. This could create the perception of an external and imposed project, rather than one born from shared effort and commitment to their future [51].

The following paragraphs will analyse the methodology and strategies suggested to the local communities under study to create a foundation for the development and achievement of a Smart Village, based on a concrete experience [52]. Social entrepreneurship, when approached consciously and with the premise of sharing not only strategies but understood as a process that helps the community and, first and foremost, provides work opportunities in small rural areas, represents one of the most powerful levers to stimulate local participation and promote change in rural areas of Italy and Europe. It aims to foster commitment to social well-being and the resolution of community problems, positioning itself as an alternative to the mere pursuit of profit. Initiatives of this kind contribute to the creation of social, cultural, and environmental value, significantly improving the quality of life in local communities. Social enterprises, often managed by cooperatives, associations, or foundations, play a crucial role in local development through activities ranging from the production of goods and services to the management of common spaces and the promotion of cultural and social projects [53]. These organizations not only generate employment but also strengthen social cohesion and encourage active citizen participation, with citizens becoming true organizers of the community's operations [54]. This community-based activity, rooted in the territory, relies on the ability to respond to local needs through sustainable and inclusive business models. Two concrete examples illustrate these principles: the first involves the *Verbumcaudo* farm in the Madonie, acquired by the Sicilian Region and managed by the *Verbumcaudo* social cooperative, currently being redeveloped with funds from the National Rehabilitation and Resilience Plan (PNRR) for the creation of multimedia spaces, workshops for schools and youth, and energy requalification, with an investment of 5.3 million euros. The second case is the Borgo di *Canistro*, financed by the PNRR with 1.6 million euros for the "Dimension *CRESCITA*" project, which includes the rehabilitation of a deconsecrated church as a Cultural Hub, a building as an InfoPoint and environmental education centre, and the creation of an Innovation Hub for the environment. The adoption of modern technologies such as Li-Fi and touchless totems enhances the local heritage, supporting slow and sustainable tourism. These examples demonstrate how social innovation and local participation, within social enterprises, can significantly contribute to the enhancement and revitalization of territories. It is essential that such initiatives respond to local resources and actively involve the community in the decision-making process. Collaboration among different communities and the sharing of resources and knowledge represent a valuable contribution to the growth and well-being of inner areas, promoting a sense of belonging and cooperation. For a EU Project developed on this issue, please refears to *SIAS2—Sicilian Inner Areas Summer School*: Project No. 2024-1-IT03-ESC30-SOL-000228603.

3.1.2 Smart Governance

Governance plays a crucial role in *Smart Villages* [55], complementing the fundamental role of the community and integrating technology, participation, and sustainability to make public policies more inclusive, transparent, and effective. In a smart approach, administrations must be able to seize the opportunities offered by funding programmes to attract resources for local development. This approach is not limited to project planning but requires the integration of the population, which is essential for active participation in decision-making processes. The use of digital portals is one of the main tools to encourage community engagement [56]. These portals provide citizens with the opportunity to express opinions, participate in surveys, stay informed about local initiatives, and actively contribute to the design of public policies. In this way, participation is no longer confined to traditional forms of interaction but becomes continuous and accessible, even for those residing in remote areas. Digitalisation allows local administrations to gather real-time *feedback*, thus improving the quality of decisions and making processes more democratic and transparent. Digital platforms are important not only in the participatory phase but also for systematising projects, proposals, and solutions aimed at promoting initiatives to be launched.

A notable example of this approach can be found in the *smart village* initiatives in Spain [57], where some administrations in mountain regions have promoted a balance between technological modernisation and the preservation of local culture. Here, *governance* has facilitated the digitalisation of public services and created digital platforms that connect citizens with local institutions [58]. The approach focuses on consultation and *co-design* with the community, enhancing the active role of citizens in decisions related to the territory and technological innovation. The case of Valencia also demonstrates how effective management requires considering the representation of interests related to sustainability and active citizenship. This approach sees rural areas as an integral part of a strategic plan involving the entire community, valuing rurality not as a periphery but as a fundamental resource [59].

In *smart governance* [60], digitalisation is a means of collecting and analysing data related to cultural, architectural, and natural heritage, creating databases accessible to all territorial actors for new projects. Some *governance* assessments might consider digitalisation, such as H-BIM for heritage, as done in Veberöd, Sweden, not only to start monitoring the heritage but also to track it and assess the impacts that certain technological solutions could undergo if integrated. Heritage mapping, in fact, allows for identifying underutilised resources and planning rehabilitation and restoration interventions that respect the local identity. Understanding the condition of buildings—whether completely outdated, partially recoverable, or already restored—enables more efficient heritage management, combining cultural enhancement, sustainability, and innovation. Technologies such as H-BIM, which catalogue heritage based on its historical significance, condition, and energy efficiency, provide an essential tool for managing and monitoring cultural and natural assets [61].

Energy efficiency represents another crucial aspect of *smart governance*, applicable to both operational and disused buildings. The adoption of advanced technologies enables monitoring energy consumption and identifying areas of inefficiency, reducing consumption and optimising resources to benefit the environment and the local economy. Specifically, energy retrofitting of disused buildings, through participation in funding programmes, can transform unused spaces into resources for the community, with a positive impact on efficiency and sustainability.

In this context, energy communities represent a sustainable development model that reduces external energy dependence and promotes local autonomy [62]. By bringing together citizens and businesses to produce and share renewable energy, energy communities promote efficiency and sustainability while offering new economic opportunities for the regions. These initiatives, often involving partnerships between local authorities, universities, and the private sector, are essential for implementing innovative solutions, improving service quality, and raising community awareness of responsible resource use. A significant example can be found in Finland, where some rural villages have implemented a digital *governance* system that integrates information flows, energy resources, and tourism activities. This approach, characterised by close collaboration between the public and private sectors, creates an ecosystem capable of improving the quality of life for residents while respecting environmental heritage. Access to European funds [63] represents a fundamental opportunity to finance *smart governance* initiatives related to digitalisation, energy efficiency, heritage enhancement, and civic participation. However, to make the most of these resources, local administrations must have the necessary skills to draft effective funding proposals [64]. This enables the implementation of innovative projects that improve the quality of services provided to citizens and promote a sustainable economic development model. Furthermore, it is essential to highlight the importance of effective communication between *governance* and citizens. Broader dialogue fosters a sense of participation and facilitates the promotion of initiatives. Greater interaction implies closer collaboration and a higher willingness to undertake common activities, contributing to the perception of *governance* not as distant but as a reality close to and integrated into the lives of citizens. *Smart governance* is based on the strategic use of technology to enhance the efficiency, participation, and sustainability of public policies. The digitalisation of decision-making processes, heritage enhancement, energy monitoring, energy communities, and access to European funds represent key tools for creating an inclusive and innovative territorial management model. This combination of elements fosters the development of resilient communities, ready to face contemporary challenges and build a sustainable future.

3.1.3 Rehabilitation and Enhancement of Tangible and Intangible Heritage

If the community and *governance* play a decisive role, architectural heritage, although not considered an active element of the territory, passively constitutes one of the fundamental aspects to be considered [65]. Heritage, both tangible and intangible, is a key resource for local communities [66], as it represents not only the historical and cultural memory of a territory but also a potential tool for economic and social development [67]. Regarding intangible heritage, digitisation emerges as one of the most powerful tools for enhancement and management, as it allows for the collection, preservation, and dissemination of historical, artistic, and cultural information in a manner that is accessible and usable by a global audience. The creation of digital archives, virtual museums [68], and interactive platforms, as well as the ability to experience heritage through virtual and augmented reality, not only preserves it but also makes it accessible in innovative ways, expanding the reach to a broader and more diverse audience [69]. The digitisation of local traditions, crafts, and oral languages contributes to preserving and promoting cultural identity by collecting testimonies and traditions that might otherwise be lost.

The promotion of events, festivals, and fairs related to local tradition and culture supports this process and represents another crucial aspect for the enhancement of intangible heritage, particularly for slow tourism [70]. Such events can, in fact, not only enrich the town but also create opportunities for vitality and rediscovery of local traditions. These events, which celebrate gastronomic, artisanal, and folkloric traditions, provide unique opportunities to engage residents and attract visitors who are sensitive to authentic culture and sustainable tourism. Through careful and targeted management of these events, communities can stimulate the local economy, promote active citizen participation, and raise awareness of a type of tourism that is more respectful of the environment and local traditions [71]. Furthermore, slow tourism [72], as mentioned in the introduction to this chapter, allows for the in-depth discovery and appreciation of the distinctive features of a territory, showcasing its territorial vocation, reducing the impact of mass tourism, and offering more authentic and immersive experiences. This is why tangible and intangible heritage are in perfect synergy; popular and religious events could, in fact, trigger virtuous mechanisms for the rehabilitation and enhancement of the built heritage.

As for tangible heritage, the protection and enhancement of the built heritage require a more technical and structured approach that can ensure the conservation of architectural heritage without compromising its historical and cultural value [73]. To this end, the adoption of manuals, guidelines, atlases, and practice codes becomes essential for managing traditional architecture and historically significant architectural sites, such as well-preserved small urban contexts [74]. First and foremost, this allows for understanding how to intervene while simultaneously spreading and promoting sustainable and compatible rehabilitation practices. These tools provide precise guidelines on how to intervene in a conservative and respectful manner, promoting rehabilitation interventions that are both effective and compatible with sustainability [75]. This is what is foreseen by the latest law of the Sicilian Region, which requires that the new PRGs to be approved in the next five years include supporting documents for interventions on heritage. The conservation of the built heritage is not merely an aesthetic matter but one of respect for the historical, architectural, and cultural identity of a place, and must therefore be guided by principles of good practice that are consistent with protection regulations and community needs.

Strategies for repurposing existing heritage represent another area of great importance, especially in the context of sustainability and social innovation. If *governance* and the community are aware of the potential of the territory, they can also be equally proactive in managing resources consciously, evaluating the opportunity to give new life to them [76]. Rebuilding, recovering, and giving new functions to these resources can create new work opportunities for the area. In this context, projects such as ecovillages [77], eco-sustainable accommodation structures, and co-housing for those over 65 are examples of how built heritage can be creatively reused, maintaining its original functionality but adapting it to modern needs. Ecovillages, for instance, offer housing solutions based on ecological principles, creating self-sufficient communities that promote the use of renewable energy and responsible management of natural resources, as well as the practice of rehabilitation by emulating local construction techniques and promoting traditional buildings. Likewise, co-housing for those over 65 responds to a growing social need, offering shared living spaces that promote intergenerational integration, mutual care, skill exchange, and, ultimately, the well-being of elderly citizens in these small contexts.

Finally, creative and *innovation hubs* for the local community play a crucial role in fostering economic growth, creativity, and social innovation. These spaces, which can host artists, start-ups, and social enterprises, provide opportunities for training, networking, and development for those wishing to launch innovative or cultural projects. By creating innovation centres and incubators for ideas, local communities can stimulate the production of added value in key sectors such as digital craftsmanship, sustainability, and technology, with a positive impact on the economic and social fabric, while also developing strategies and pathways for the enhancement of the territory. Creative hubs not only promote employment and integration but also contribute to a cultural and social transformation, where the community becomes actively involved in defining and realising its future development. The enhancement of tangible and intangible heritage through digitisation, the promotion of cultural events, the protection of built heritage, reuse strategies, and the creation of creative hubs represents a set of practices that significantly contribute to the sustainability and growth of local communities. These approaches, when integrated into a coherent territorial development plan, not only preserve and promote cultural identity but also stimulate social and economic innovation, creating a balance between tradition, modernity, and sustainability. In the context of the thesis, a useful approach to promoting the rehabilitation and enhancement of local architecture was the development of a tool such as the Guidelines, designed to support the conservation of built heritage and to promote knowledge of tools, technological solutions, and innovative methodologies applicable to the local heritage of the Municipality of Polizzi Generosa. This tool aims to raise citizens' awareness of an integrated development model, based on the collection and systematisation of past activities and research to enhance traditional architecture and encourage a conscious and sustainable approach to territory rehabilitation.

The Guidelines represent a reference to guide interventions compatible with the peculiarities of rural and urban contexts, considering material, cultural, and landscape specifics. Moreover, through the adoption of new technologies and the promotion of collaborative practices between local communities and administrations, the tool aims to enhance the territory's capacity to adapt to changes while preserving the authenticity and cultural identity of the places. A central aspect of this approach is the active involvement of local communities, encouraging direct participation in decision-making processes and fostering constructive dialogue between experts, administrators, and citizens. This dialogue not only promotes greater awareness of the value of built heritage but also helps define shared strategies for its rehabilitation, safeguarding, and enhancement, with a focus on sustainability and innovation. Through the Guidelines, which will be addressed in the following section, it is hoped to stimulate an integrated management model capable of combining tradition and modernity, transforming local heritage into a dynamic resource for the economic, cultural, and social development of the communities.

3.1.4 Technology and Digitalisation

Technology and digitalisation are key elements for the enhancement, management, and sustainable development of communities, particularly in rural areas and small towns. The adoption of digital tools not only facilitates access to information and services but also opens new opportunities for integrating cultural heritage, energy sustainability, and improving quality of life. Apps and websites are among the main tools for promoting cultural heritage, providing a direct communication channel with visitors and enabling them to explore history, traditions, natural beauty, and local specifics in an interactive and engaging way. Thanks to digitalisation, it is possible to create platforms that allow tourists and residents to discover both tangible and intangible heritage in an innovative manner, with virtual tours, multimedia content, and up-to-date information, fostering dynamic and participatory management [78].

Dedicated apps and websites can also integrate useful functionalities for the community, such as mapping cultural routes, organising events, and promoting sustainable tourism activities, thereby contributing to a richer and more interactive experience [79]. The experience initiated by the Municipality of Polizzi Generosa, through the portal visitpolizzigenerosa.it, demonstrates that it is possible to systematise resources, ensuring knowledge of the places, experiences, and excellences available in the territories. This work has been partially carried out in some municipalities of the area, but it deserves further development and expansion to involve other municipalities in the participation and systematisation of resources. The work already initiated and expanded can help the entire sector share and systematise what it has to offer. Additionally, the use of advanced technologies, such as augmented reality, would enrich visitor experiences [80] with real-time information and interactive content, providing tourists and residents with an immersive experience that connects them with the history and culture of the territory [81].

Another fundamental aspect related to technology and digitalisation is the management of energy resources, particularly through the creation of digital energy communities. These communities allow for optimising local energy use by real-time monitoring of consumption, enabling the management of energy flows and the adoption of sustainable solutions such as solar power, energy efficiency, and collective self-consumption. Digital technologies play a crucial role in ensuring that resources are managed efficiently, and users can autonomously and consciously monitor and control their energy consumption. The adoption of digital platforms for energy efficiency monitoring allows for collecting valuable data, analysing consumption patterns, and implementing solutions to reduce waste, contributing to more sustainable management of local resources. At the same time, digital technologies [82] can significantly improve urban aesthetics, particularly through the implementation of modern infrastructure such as public Wi-Fi and fibre optic networks [83]. Public Wi-Fi and fibre optics are also essential for facilitating access to tourist, cultural, and administrative information, creating a connection between citizens, visitors, and local institutions [84]. Creating a digital connection network accessible to all not only improves the quality of life for residents and visitors but also stimulates the development of economic activities linked to innovation and creativity. It also ensures the internationalisation of local specialties, especially products that could have a market globally. Rural areas and small towns, which often suffer from technological isolation, can benefit from these solutions to reduce the digital divide, ensure access to online services, promote social inclusion, and attract new entrepreneurial initiatives. Finally, the centralisation and management of resources and services through "control rooms" [85] represent an advanced solution for real-time monitoring and management of local operations. Digital control rooms can centralise data from various systems (energy, traffic, security, public services) and provide crucial information for swift and informed decision-making. The centralisation of data allows for more efficient resource management and better coordination of daily activities such as infrastructure maintenance, emergency management, urban planning, and public safety. Control rooms can also integrate artificial intelligence systems and predictive analytics, improving the ability to anticipate and solve problems before they become critical, ensuring a timely and effective response [86]. Given the region's high seismic risk, this enables real-time monitoring of potential hydrogeological instability and promotes preventive interventions, aiming at predictive maintenance of resources and assessing interventions to support. The integration of technology and digitalisation in heritage management practices, energy resources, and urban aesthetics offers significant opportunities for the sustainable development of local communities [87]. The use of apps and websites, the creation of digital energy communities, the implementation of advanced technological infrastructure like Wi-Fi and fibre optics, and the centralised management through control rooms are all solutions that improve quality of life, foster innovation, optimise resources, and ensure a more sustainable and inclusive future for the territories [88]. Technological innovation is a cornerstone for the progress and development of contemporary societies, including the most remote and disadvantaged areas, such as inner regions. This concept refers not only to the introduction of new technologies but also to the improvement of existing technologies, with the goal of optimizing processes, products, and services [89]. It is not limited to the creation of new devices or software; it also includes the implementation of innovative methods for using already existing technologies to solve complex problems and improve the quality of life, local productivity, and the daily activities of a community focused on progress [90]. In inner areas where resources are often limited and infrastructural challenges are significant, technological innovation can play a crucial role in overcoming historical barriers and promoting sustainable development. It allows for addressing issues such as geographic isolation, lack of essential services, and economic marginalisation. Through the adoption of advanced technologies, connectivity can be improved, access to public services facilitated, and new economic opportunities created. Advanced technologies, in fact, facilitate access to essential services such as healthcare, education, and security, significantly improving the quality of life in inner areas. For example, digital technologies allow residents to access high-quality healthcare services via telemedicine or participate in online educational programs, including training [91]. Technological innovation stimulates economic growth through the creation of new job opportunities and support for the development of local businesses. Furthermore, it promotes environmental sustainability through the adoption of *green* technologies, such as renewable energy, and the optimisation of natural resource use, as well as the practice of using new materials from local waste for the rehabilitation of the built heritage [16]. The adoption of innovative technologies can help bridge the gap between urban and rural areas, improving access to services and opportunities more equitably [92]. The spread of high-speed internet, for instance, can connect isolated communities to the rest of the world, facilitating trade, education, and civic participation. It has the potential to profoundly transform the social and cultural dynamics of communities, fostering greater participation and social cohesion. The digitalisation of public services, for example, can make local administrations more transparent and efficient, strengthening citizens' trust in institutions. Despite the numerous benefits, technological innovation also presents significant challenges. It is essential to ensure that new technologies are accessible to everyone, avoiding the creation of new forms of exclusion, particularly towards older generations. Therefore, the digitalisation process must also consider training processes and the provision of appropriate technologies to guarantee access and usability for all. Technological innovation represents a powerful tool for transforming inner areas, improving quality of life, promoting economic growth, and reducing inequalities. However, its success depends on careful planning and the inclusion of local communities at every stage of the innovation process.

First and foremost, the training process is crucial. The use of technology in remote contexts like these is not easily applicable, both for young people and the elderly. Projects like *NONNO Lab*, *Nonni su Internet*, and *Terza Età Online* are just a few examples that have been undertaken with great success in Italy. Also noteworthy is the proliferation of digital help desks established in small municipalities, aimed at providing free consultations to facilitate access to the Internet and ensure the use of SPID, an initiative that is now expanding rapidly to other municipalities and parts of the world. This process, based on the adoption of advanced digital tools and platforms, has revolutionized the way information is generated, processed, distributed, and used, deeply influencing the economy, culture, and functioning of both public and private institutions [93]. Digital innovation is not limited to the mere introduction of new technologies but also involves a profound cultural, organizational, and methodological change. To fully exploit the potential of emerging technologies, organizations must rethink their operational models, revise inner processes, and develop new skills. This transformation journey, also proposed by the EU in terms of an evaluation matrix and key phases in digital transformation, requires a consistent and strategic integration of digital technologies, addressing challenges related to security, data management, and sustainability. In small communities, where services often depend on larger urban centres, the digitalisation of administrative practices becomes an unavoidable priority. The implementation of digital help desks simplifies and speeds up bureaucratic processes, reducing the distance between citizens and decentralized administrations and decreasing dependence on larger cities [94]. A robust network infrastructure, starting with broadband, is just the first step. Next, it is essential to invest in the use of databases and integrated platforms that support the management and secure sharing of data; digital clouds and protection measures for these data are also crucial. This digitalisation not only concerns the management of administrative practices but also has a direct impact on the quality of services offered to the community. Data, in fact, become a valuable tool to inform citizens in real-time, for instance, about traffic conditions or road closures [43], to promote safe and informed mobility in mountain areas, and to encourage the use of public transport [95]. In this way, digital innovation contributes to building a more efficient and connected community, strengthening participation and quality of life even in the most remote areas.

3.1.4.1 Accessibility and Design for All

The rural and mountainous areas are often characterised by complex topography and constraints related to the preservation of historical-architectural heritage, making the implementation of accessibility interventions even more challenging. While the previous section discussed digital accessibility, it is equally important to consider physical accessibility. Ensuring universal access to technologies, goods, services, and infrastructure is essential to promote inclusion and territorial equity, while also supporting the enhancement of these areas [96]. The concept of "*Design for All*" presents itself as a design philosophy that embraces an inclusive vision, aimed at creating environments, products, and services usable by anyone, regardless of their physical or cognitive abilities. It was introduced by architect Ronald L. Mace in 1985 to describe a design process that responds to the needs of a wide range of people, including those living with disabilities [97]. Over the years, the concept has evolved through various approaches, such as *Inclusive Design* and Human-Centered Design, eventually leading to *Design for All*, as established by the 2004 Stockholm Declaration. The goal of these approaches is clear: to design for human diversity.

In 1997, the Center for Universal Design at the University of North Carolina formalised seven principles that guide *inclusive design* [98], which, to this day, must remain the core of any design process, both before and after:

- equitable use: the use must be equitable, allowing everyone to access the same opportunities without discrimination;
- flexibility in use: the design must ensure flexibility, adapting to the different abilities and preferences of users;
- simple and intuitive: products and environments must be simple and intuitive to use, without requiring complex instructions;

- perceptible information: information must be conveyed clearly, regardless of environmental conditions or the user's sensory abilities;
- tolerance for error: the design must minimise the risks of error or inappropriate use;
- low physical effort: the use must require minimal physical effort;
- size and space for approach and use: spaces must be suitable for access and use by everyone.

In these contexts characterised by a high historical and scenic value, the adoption of these principles must be integrated with respect for the existing urban fabric and the need to preserve the authenticity of the places. An inclusive project in the inner areas of Sicily, therefore, must not only address the needs of the resident population (which often includes mainly elderly people and those with disabilities), but also do so by considering geographical specifics and the cultural, and especially architectural, pre-existing elements. Physical accessibility relates to the design and adaptation of public and private spaces to ensure they can be used by everyone, including people with disabilities [97]. To improve physical accessibility, it is essential to implement solutions such as suitable ramps and elevators that allow access to buildings, structures, and parts of urban areas without barriers. Ramps must be designed with an appropriate slope, sometimes even partially replaced by small stairways when too steep, and elevators should be spacious and easy to use, integrating into the urban context, even if this sometimes goes against traditional principles of urban decor. For this reason, knowledge of traditional architecture and techniques is useful to understand the type of interventions to be implemented, given the numerous architectural barriers found in the many staircases and kerbs of the inner villages [45]. This accessibility must also be evaluated within buildings, contributing to the proper usability of spaces for those with disabilities.

Along with physical accessibility, visual accessibility is equally important. The design must be simple and intuitive, ensuring that it is easily understandable and usable without the need for complicated instructions. Additionally, the integration of assistive technologies is essential to allow people with disabilities to fully participate in society. These tools include vocal screens, screen readers, and other aids that enhance interaction with the digital and physical world; the use of fonts suitable for easier reading of texts and signage is also crucial. Finally, user involvement in the design process is essential. Moreover, signage, whether monumental or street, must be clear and highly visible, with appropriate contrasts and large characters, and include tactile and Braille indications for visually impaired people. The flooring and furnishings should be designed to ensure safety and comfort while adhering to principles of urban decor as outlined by possible colour plans, rehabilitation manuals, or guidelines, such as those for the Municipality of Polizzi Generosa, output of the following thesis. It is important to use non-slip materials and avoid obstacles that may impede movement. The experiences and *feedback* from people with disabilities provide valuable information to refine and improve accessibility.

The application of these principles extends to various sectors, such as architecture and urban planning, by designing buildings and public spaces with ramps, elevators, and accessible pathways. Industrial and user interface design also benefits from these principles, with products and services designed to be usable by a wide range of people, not least considering that the user experience must be guaranteed in every case. The design of *green* spaces should also take accessibility into account, including paths and equipment usable by everyone, such as benches and adjustable tables, as well as shaded areas for hot and humid summers or shelters for particularly cold winters.

Digital accessibility, already widely discussed in the previous section, must be integrated with physical accessibility, as done for the totems for the Municipality of Polizzi Generosa. Developed with the great synergy of CoopCulture, it allows the municipality to be known online via a QR code and expands the local cultural offerings. The initiative is aimed at both physical and digital accessibility, or rather, an innovative approach to digital heritage. Corten steel totems that enhance heritage while seamlessly integrating into the local context also respect all types of disabilities, whether visual or physical, allowing access even for wheelchair users or those with limited vision. Raising awareness and training at this stage are crucial to promote a culture of inclusion and ensure that design practices adhere to accessibility principles [99]. Training programmes aimed at designers, architects, public administrators, and citizens help spread the knowledge and skills necessary to implement universal design. Public awareness campaigns can educate the public on the importance of accessibility and the benefits it brings to the entire community.

It is also helpful to collaborate with associations representing people with disabilities to obtain *feedback* and suggestions that can improve design and inclusion practices. Accessibility regulations provide guidelines and legal requirements to ensure that buildings, public spaces, and technologies are accessible to all. In Italy, Law 104/1992 establishes the fundamental principles for the assistance and inclusion of people with disabilities, while Law 246/2005 defines specific requirements for the accessibility of spaces and structures. At the European level, Directive 2016/2102 sets standards for the accessibility of websites and mobile applications of public authorities. To ensure compliance with these regulations, it is important to keep them updated to reflect new technologies and emerging needs. Implementing these practices not only improves the quality of life for people with disabilities but also enriches society by promoting inclusion and equality.

3.2 Tools for Managing Smart Villages

In the development of *Smart Villages*, either by analogy or contrast, it is essential to follow a structured pathway, moving from strategic elements to operational ones. Among the key aspects to consider is the identification of the main *stakeholders*: *governance*, community, heritage preservation, and the implementation of digital technologies, as discussed in the previous sections. These *stakeholders*, however, require a precise process to follow, organised into various stages that include guidelines [100], forming the strategic foundation, followed by *best practices*, *business plans*, and finally management protocols. In detail:

- guidelines: provide a framework of general principles and strategic orientations capable of steering the project towards a digital and sustainable approach tailored to the territorial specificities and the needs of communities;
- best practices: follow the guidelines as proven and replicable operational models, offering concrete solutions inspired by successfully implemented experiences. They help translate general principles into practical examples that can be locally adapted;
- business plan: this represents the project's economic and financial framework. Based on the guidelines and *best practices*, it details economic sustainability and the necessary investment plans, ensuring that resources are efficiently allocated to support the project's long-term objectives;
- management protocol: This outlines operational procedures, roles, and responsibilities for each project phase. It guides the practical execution of actions, addressing *governance*, community integration, heritage preservation, and the implementation of digital technologies.

Each element builds a support system necessary for the next, ensuring that the project develops with a solid strategic foundation, tested operational models, a well-defined financial plan, and an effective management framework.

3.2.1 Guidelines and Best Practices

Considering the analysis conducted so far, it is appropriate to examine thematic maps of rural areas to understand the current situation across different contexts. These maps aim to identify municipalities or local communities that can be considered exemplary in terms of project development and to assess how initiatives can be re-evaluated to promote active participation from communities that have remained uninvolved [101]. The maps consider the total number of municipalities or rural communities and trace the paths they have followed in recent years to implement projects aimed at creating Smart Villages [102]. Following specific processes is essential for the shared management of resources, the development of strategies [103] and synergistic projects. This can be achieved through guidelines and best practices [104]. The subsequent list of actions outlines strategies to foster the development of Smart Villages in rural areas. Starting from general guidelines [105], best practices are derived, specifying how these should be applied according to key themes:

Community integration and active participation: Engage citizens in decision-making and the management of cultural assets to foster an aware and responsible community.—*Best practices*: establish working groups and listening platforms to gather proposals and *feedback* from citizens; organise *co-design* sessions, workshops, and focus groups to encourage active participation and build shared solutions; promote initiatives for associationism and community cooperation, encouraging the creation of local cooperatives for managing services and heritage.

Inclusive and transparent governance: Implement *governance* that is transparent and capable of addressing community needs, with clear policies for the protection and enhancement of heritage.—*Best practices*: develop a strategic plan that integrates conservation, innovation, and sustainability, ensuring it is accessible to all citizens; facilitate public–private partnerships to attract resources and expertise, supporting cultural and heritage preservation initiatives; establish monitoring indicators to periodically assess the impact of preservation initiatives and implement improvements based on the results; integrate tools for heritage management.

Technology implementation and digitalisation: Leverage technological tools to enhance access to information, communication, and heritage conservation, as well as to optimise local services.—*Best practices*: digitise cultural assets and create an online platform allowing both citizens and tourists to explore the cultural and natural heritage of the village; use sensors and monitoring technologies for predictive maintenance of historical assets, preventing damage and preserving original materials and structures; develop apps or digital portals to engage the community in decision-making and collect *feedback* or suggestions on ongoing initiatives.

Enhancement of tangible and intangible Heritage: Safeguard the village's tangible and intangible resources and promote them as cultural and tourist assets.—*Best practices*: undertake restoration and maintenance projects that respect traditional construction techniques and local materials, preserving the authenticity and historical value of assets; organise festivals, exhibitions, and cultural trails that showcase local traditions and skills, involving artisans, artists, and the community; support the transmission of traditional knowledge and local expertise by offering courses and workshops to educate new generations and visitors.

Sustainability and circular economy: Adopt a circular economy model to promote environmental sustainability and waste reduction, encouraging recycling and reuse in local activities.—*Best practices*: encourage local businesses to use renewable resources and create products with sustainable materials; organise awareness campaigns on the importance of sustainability and the circular and responsible management of resources.

Accessibility and inclusion: Ensure that initiatives and points of interest are accessible to everyone, regardless of physical abilities or cultural background.—*Best practices*: ensure that cultural trails, buildings, and digital information are accessible, and provide structural adjustments to venues for inclusion.

The development of *Smart Villages*, therefore, requires a holistic and collaborative approach. The guidelines and *best practices* outlined above are essential to ensure that such projects are sustainable, inclusive, and capable of improving the quality of life in rural communities, while also implementing proposals and projects for and with the community. Technological innovation, when driven by community engagement and a long-term vision, can transform villages into centres of prosperity and well-being. These are, therefore, fundamental and necessary references to guide the design and management of *Smart Villages*; regulatory documents that provide guidance on how to address specific aspects of the management and development of these villages, offering a methodological and operational framework. *Best practices*, on the other hand, represent the most effective strategies and solutions, developed based on positive experiences and tangible successes. These tools are crucial to ensuring that projects align with the highest standards and efficiently and sustainably meet local needs.

3.2.1.1 Application Areas

The guidelines and *best practices* identified are useful for developing management and organizational processes for *Smart Villages*, drawing inspiration from both the Atlas developed and the concrete needs identified in the territory. Each area of application requires a targeted approach to ensure optimal results, also considering the principles of the 2030 Agenda for Sustainable Development. For this reason, the Guidelines refer to the 12 categories of the developed Atlas, while also covering additional areas, sometimes even intersecting, to contribute to the sustainable and *green* development of the rural areas. Among the main areas of intervention are the management of technological infrastructure, environmental sustainability, economic and social development, and public safety and health. As an example, these may concern technological infrastructure, including the design of connectivity networks and the integration of *IoT* devices to facilitate land management, as well as monitoring and predicting maintenance interventions. These guidelines may also promote the adoption of renewable energy and advanced waste management practices. In the economic and social context, the focus is also on active community involvement and the promotion of initiatives related to sustainable tourism. In the context of public safety and health, they may involve the introduction of telemedicine systems and tools for monitoring air quality, thus contributing to the well-being of the population.

3.2.1.2 Objectives and Benefits

The main objectives are to ensure effective and efficient land management, promoting an integrated approach that combines technological innovation, environmental sustainability, and community participation. The benefits include the definition of quality standards and operational procedures that encourage the adoption of advanced technologies and strategies consistent with the principles of sustainable development. These include improving territorial organization, optimizing available resources, and promoting inclusive and participatory *governance*. Among the main benefits are the improvement of the quality of life and the consistency of services provided, enabling more transparent and accessible management for all inhabitants. The adoption of optimized solutions significantly reduces operational costs, freeing up resources for further strategic investments. Increased environmental sustainability, achieved through innovative and low-impact practices, contributes to the protection of the territory and the enhancement of quality of life. Another benefit is the strengthening of social cohesion, thanks to processes that promote inclusion and active participation of the local community. This involvement not only values the role of residents but also stimulates the creation of new economic and social opportunities, consolidating a sense of belonging to the territory. The standardization of practices facilitates the replicability of virtuous models, allowing the spread of successful experiences in other contexts, with a positive impact on a larger scale.

3.2.1.3 Development and Implementation

Their development and implementation are based on a structured and systematic process, considering the specific characteristics of the territory and the challenges that need to be addressed. The first phase consists of a detailed analysis of local needs, aimed at identifying priorities, critical issues, and opportunities. This preliminary study provides a clear and contextualized framework, essential for developing guidelines and operational practices that effectively respond to the identified needs. The drafting of the guidelines represents the next step and is characterized by the formulation of recommendations based on the conducted analysis. These recommendations include innovative technical and operational solutions that reflect the principles of sustainability, efficiency, and inclusiveness. The goal is to define practical tools and strategies that can optimize resource management and promote local development.

Implementation requires active involvement of *stakeholders* through targeted training programs, designed to ensure a deep understanding of the suggested practices. At the same time, the adoption of the guidelines must be supported by a continuous monitoring system, enabling the tracking of progress, identification of potential issues, and the introduction of modifications or adjustments when necessary.

A crucial element is the *feedback* and periodic review mechanism, which ensures the constant relevance and potential modification or optimization of the guidelines over time. This iterative approach allows for prompt responses to new needs or changes in the context, ensuring flexible management and a focus on continuous improvement.

3.2.1.4 Role in Innovation and Change

Guidelines and *best practices* play a crucial role in promoting innovation and territorial change for achieving the goals within *Smart Villages*, acting as catalysts for the development of advanced and sustainable solutions. A structured and reliable foundation facilitates the introduction of emerging technologies and the integration of innovative methodologies, enhancing the ability of local communities to adapt to future changes and challenges (Fig. 3.3). By promoting advanced technologies and adopting innovation-oriented strategies, the guidelines foster the modernization of infrastructure and services, stimulating a profound transformation of rural communities. This process helps make them models of sustainable development, capable of combining tradition and progress in a virtuous balance. *Best practices* facilitate change through the standardization and sharing of effective solutions, creating a favourable environment for experimentation and the spread of new ideas. This approach encourages a continuous cycle of improvement, promoting not only the application of innovative technologies but also the development of collaborative and participatory strategies, which strengthen the resilience and competitiveness of rural communities.

3.2.2 Business Plans

The effective management of a *Smart Village* requires the creation of well-structured management protocols and strategic *business plans*. Management protocols establish the necessary standards for implementing and maintaining smart and sustainable technological solutions, while the *business plan* provides a strategic and operational vision to ensure the economic sustainability and long-term success of the solutions to be undertaken. This section will explore the key elements of management protocols and *business plans*, discuss the implementation and monitoring phases, and highlighting the importance of an integrated approach to ensure the success of *Smart Villages*.

3.2.2.1 Key Elements of Business Plans

In the context of sustainable development in rural areas, business plans play a crucial strategic role in translating visions and strategies into concrete and measurable actions. The key elements of a business plan for a Smart Village are structured across several interconnected levels, addressing the specific characteristics of the rural context under study. A primary element is the enhancement of cultural and natural heritage, which in rural areas can include historical centers, artisanal and gastronomic traditions, and unique ecosystems. These heritages represent the core of the project and the foundation on which initiatives that attract both residents and visitors should be built. A second element is the involvement of the local community, necessary to ensure the success of the project and promote participatory governance. The community becomes a central actor in managing and promoting the territory through cooperatives, associations, and co-design processes. Digitalisation and technological innovation form another pillar. In rural contexts, digital transformation could include creating digital archives of cultural and natural heritage, using applications for tourism and local service management, and implementing monitoring technologies to safeguard environmental and architectural assets. Finally, financial analysis and an economic sustainability plan complete the business plan. Considering the often-fragile nature of the local economic fabric in rural areas, it is essential to identify diversified resources, from European and national rural development funds

COMMUNITY INTEGRATION AND ACTIVE PARTICIPATION

Engage citizens in decisions and the management of cultural heritage to foster an informed and responsible community:

- Create working groups and listening platforms to collect proposals and feedback from citizens.
- Organize co-design sessions and workshops to encourage active participation and build shared solutions.
- Promote associationism and community cooperation initiatives, encouraging the creation of local cooperatives for the management of services and heritage.

INCLUSIVE AND TRANSPARENT GOVERNANCE

Implement transparent governance capable of responding to the community's needs, with clear policies for the protection and enhancement of heritage.

- Develop a strategic plan that integrates conservation, innovation, and sustainability, and is accessible to all citizens.
- Facilitate public-private partnerships to attract resources and expertise, supporting cultural initiatives and heritage protection.
- Establish monitoring indicators to periodically assess the impact of protection initiatives and make improvements based on the results obtained.

TECHNOLOGY IMPLEMENTATION AND DIGITALIZATION

Use technological tools to improve access to information, communication, and heritage preservation, as well as to optimize local services.

- Digitize cultural assets and create an online platform that allows both citizens and tourists to explore the cultural and natural heritage of the village.
- Use sensors and monitoring technologies for predictive maintenance of historical assets, preventing damage and preserving original materials and structures.
- Create apps or digital participation portals to engage the community in decision-making and gather reports or suggestions on ongoing initiatives.

VALORIZATION OF MATERIAL AND IMMATERIAL HERITAGE

Safeguard the material and immaterial resources of the village and promote them as cultural and tourist assets.

- Carry out restoration and maintenance interventions that respect traditional techniques and local materials, preserving the authenticity and historical value of the assets.
- Organize festivals, exhibitions, and cultural tours that highlight local traditions and skills, involving artisans, artists, and the community.
- Support the transmission of traditional knowledge and local know-how by offering courses and workshops to educate new generations and visitors.

SUSTAINABILITY AND CIRCULAR ECONOMY

Adopt a circular economy model to promote environmental sustainability and waste reduction, encouraging recycling and reuse in local activities.

- Encourage local businesses to use renewable resources and create products with sustainable materials.
- Organize awareness campaigns on the importance of sustainability and responsible resource management.

ACCESSIBILITY AND INCLUSION

Ensure that initiatives and places of interest are accessible to all, regardless of physical abilities or cultural background.

- Ensure that cultural routes and digital information are available in accessible formats and make structural adaptations in historical sites for inclusion.

Fig. 3.3 Methodology for implementing guidelines and best practices for the development of smart villages. © Image by author, 2024

to potential private contributions, as well as income generated from tourism, local cooperatives, and other initiatives to be launched.

3.2.2.2 Business Plan Structure

The *business plan* is a fundamental tool for planning, organizing, and evaluating the sustainability of a project [106]. However, when applied to the context of *Smart Villages*, it takes on characteristics that distinguish it from the classical model, while maintaining some underlying analogies. Both share a general structure articulated in key sections such as the project summary, context analysis, operational strategy, and financial planning. The main objective is to define a clear path that enables the achievement of set results by optimizing available resources and maximizing benefits for *stakeholders*. Similarly, both require a thorough context analysis to identify opportunities and challenges, as well as a detailed economic feasibility assessment. Moreover, in both cases, success depends on the ability to build strategic partnerships, which in the case of *Smart Villages* often involve local actors, public entities, and non-governmental organizations. What substantially differentiates a classical *business plan* from one developed for *Smart Villages* is the context in which it is applied and the goals it pursues. In the traditional model, the focus is often on urban or business sectors, characterized by highly competitive density and the pursuit of immediate financial profits. In rural villages, however, the *business plan* takes on a broader dimension, where economic objectives intertwine with social and environmental goals. The focus is not just economic growth, but also countering depopulation, social cohesion, and the enhancement of cultural and natural heritage. Consequently, the target changes: while the classical *business plan* primarily addresses private investors and customers, that for *Smart Villages* involves local communities, public entities, and tourists, aiming to meet collective needs through an inclusive and sustainable development model [107]. Another distinguishing element is represented by the key success factors. In the classical business plan, the focus is often on product or service innovation, competitiveness, and operational efficiency. In the case of *Smart Villages*, success depends on the ability to engage the local community, protect and enhance heritage, use technology appropriately, and create participatory *governance*. Technology, which in the traditional model is used predominantly to innovate products and services, becomes a key tool to improve the quality of life in rural areas. Initiatives such as the digitalisation of heritage, environmental monitoring, or the creation of platforms for active citizen participation are concrete examples of this approach. The economic model also differs significantly. While the classical *business plan* relies on economies of scale and maximizing return on investment, the one for *Smart Villages* integrates mixed economic models, where the circular economy, experiential tourism, and community cooperation play a central role. This reflects the nature of rural projects, which often seek to balance economic and social aspects, aiming to generate a positive impact on people's quality of life and the environment. Finally, the evaluation of impact changes. In the classical model, success is predominantly measured in terms of profit and market growth, while in the context of *Smart Villages*, the evaluation is holistic. Economic indicators are complemented by social metrics, such as the level of community cohesion or the ability to counter depopulation, and environmental ones, such as landscape preservation and reduction of ecological impact. This approach reflects a paradigm shift, where development is seen not only as an economic process but as a balance between growth, sustainability, and the enhancement of local resources. The *business plan* for a *Smart Village* is structured (Fig. 3.4) to reflect the specifics of the territory, while promoting a replicable model for other rural areas by analogy or contrast. Therefore, the following structure is proposed for consideration:

Executive Summary: A summary highlighting the role of the *Smart Village* as a driver of development for inner areas, emphasizing the benefits for residents and its cultural and tourism potential. For example, it could be presented as a "distributed laboratory" for experimenting with new models of *governance* and sustainability. Project Description: The proposed project should enhance the minor towns with interventions ranging from the rehabilitation of architectural heritage to the creation of thematic paths related to biodiversity and local history, along with the adoption of technologies that improve accessibility and public services in line with the 2030 Agenda or new goals promoted by Europe.

Project Description: The proposed project should enhance the minor towns with interventions ranging from the rehabilitation of architectural heritage to the creation of thematic paths related to biodiversity and local history, along with the adoption of technologies that improve accessibility and public services in line with the 2030 Agenda or new goals promoted by Europe.

Market Analysis: The target audience study includes examining the flow of tourists seeking authentic experiences, young entrepreneurs interested in the potential of rural areas, and residents aspiring to an improved quality of life. Industry trends highlight a growing demand for sustainable tourism and a return to smaller towns, supported by national and European policies.

Operational Strategy: The plan should include integrated activities, such as the digitization of historical heritage, with virtual tours and interactive applications; the organization of cultural festivals that highlight local traditions and flavours; the formation of cooperatives for managing social, economic, tourism, and cultural services.

Marketing Strategy: A territorial campaign could focus on the concept of "4.0" or of the future, positioning the towns as cutting-edge destinations, capable of combining authenticity and innovation for a long-term vision.

Funding Sources: The plan includes a mix of public and private resources. Key sources include European funds for digital transition, national programs for rural areas, and partnerships with companies interested in investing in *green* and innovative projects.

Financial Analysis: Analyse any cost items considered or to be considered, as well as the annual cost of the solution.

Fig. 3.4 Business plan for smart villages. © Image by author, 2025

EXECUTIVE SUMMARY

The goal of the smart village is to promote a sustainable development model that enhances local heritage, actively involves the community, and integrates advanced technologies to improve governance and services. This business plan aims to define the actions, costs, and investment potential to transform the village into a culturally active, tourist-friendly, and sustainable center, thereby attracting new economic opportunities.

PROJECT DESCRIPTIONS

The project involves an integrated approach that enhances the cultural and environmental heritage of the village, while also improving the quality of services and promoting a circular economy. The focus is on accessibility, digitalization of heritage, creation of cultural routes, and the promotion of social and sustainable initiatives.

MARKET ANALYSIS

- **Target Audience:** Local residents, cultural tourists, young entrepreneurs, rural area investors, training and research institutions, tourism operators.
- **Industry Trends:** Growing interest in sustainable tourism and immersive cultural experiences, development of technologies applied to the digitalization of villages and rural areas, and policies supporting the revitalization of small towns.

BUSINESS OBJECTIVES

- **Community and Participation:** Increase active community involvement in heritage management, with the goal of having 70% of residents participate in initiatives within the first three years.
- **Digitization and Technology:** Digitize at least 50% of the heritage within the second year, implementing monitoring technologies for structural preservation.
- **Sustainability and Tourism:** Increase tourist flow by 30% in the first three years, attracting cultural and sustainable tourists.
- **Local Employment:** Create job opportunities and form local cooperatives, reducing youth emigration and fostering the inclusion of new professions.

OPERATIONAL STRATEGY

A. Community and Participation
- Activities:
  - Organization of workshops, cultural events, and craft workshops.
  - Creation of community cooperatives for the management of cultural services.
- Costs:
  - Events and workshops: €...
  - Training and support for cooperatives: €.../year.
- Expected Results:
  - Creation of a local support network.
  - Growth in the sense of belonging and participation.

B. Governance and Partnerships
- Activities:
  - Development of local regulations for heritage protection.
  - Partnerships with private entities for co-financing cultural and technological initiatives..
- Costs:
  - Studies and strategic plans: €... one-time.
  - Partnership management: €.../year.
- Expected Results:
  - Strengthening of local governance.
  - Greater access to external funds and resources.

C. Technology and Digitization
- Activities:
  - Creation of a digital archive of heritage, accessible via an app.
  - Implementation of sensors and IoT technologies for monitoring and preventive maintenance.
- Costs:
  - Digital infrastructure and app: €...
  - Sensors and maintenance: €.../year.
- Expected Results:
  - Greater access to historical and cultural information.
  - Reduction in restoration and maintenance costs.

D. Valorizzazione del Patrimonio Materiale e Immateriale
- Activities:
  - Restoration of historic buildings, rehabilitation of basic construction, and creation of cultural routes.
  - Organization of annual events such as cultural festivals and exhibitions.
- Costs:
  - Recovery and maintenance: €.../year.
  - Promotion and tourism marketing: €.../year.
- Expected Results:
  - Increase in cultural tourism.
  - Preservation of cultural and environmental heritage.

MARKETING STRATEGY

- **Digital Promotion:** Dedicated website and app with virtual tours and reservations for guided tours.
- **Collaboration with Travel Agencies:** Agreements to include the village in cultural tourism circuits.
- **Local Awareness Campaigns:** Educating the community about the importance of heritage enhancement, with communications on social media and local media channels.

SOURCES OF FUNDING

- **European and National Funds:** Searching for funding and grants at the European level to support digital infrastructure and restoration.
- **Partnerships with Private Entities and Sponsorships:** Collaborations with companies that can fund events and activities in exchange for visibility and networking opportunities.
- **General Revenue:** Entrance fees for events, guided tours, and contributions from local cooperatives.

FINANCIAL ANALYSIS

Cost Item	Annual Cost (€)
Community events and workshops	...
Training and support for cooperatives	...
Studies and strategic plans	...
Partnership Management	...
Digital Infrastructure and Apps	...
Sensors and Maintenance	...
Rehabilitation, Restoration, and Maintenance	...
Tourism Promotion and Marketing	...
Annual Total	...

EVALUATION AND MONITORING

- **KPI:** Monitoring of community engagement, tourist flow, technology usage, and environmental impact.
- **Quarterly Reviews:** Evaluation of progress, with potential adjustments to optimize the plan and adapt to new opportunities or challenges.

Evaluation and Monitoring: The use of indicators such as the number of users, community engagement, and direct economic impact will allow progress to be measured and strategies to be adapted as the work progresses.

3.2.2.3 Project Implementation Phase

Planning and coordination: The implementation phase begins with careful operational planning that defines clear objectives, timelines, resources, and responsibilities. Once the strategic plan is outlined in the business plan, it is necessary to translate the guidelines into concrete actions. A key element of this phase is coordination among all involved actors: local administrations, communities, private businesses, public entities, and other organizations. Each partner must have a specific role, and the decision-making process should be clear, transparent, and participatory.

Compatible development of infrastructure and technologies: A fundamental part of implementation is the creation of technological infrastructures. Digital connectivity, *smart governance* solutions (such as platforms for citizen engagement), and technologies for managing natural resources (like sensors for environmental monitoring) are just some of the areas that require attention. The goal is for technology to improve the quality of life without altering the local heritage and culture but rather supporting its enhancement.

Training and community inclusion: Another crucial aspect of implementation is the training of the local community. Residents must be prepared to use new technologies and actively involved in decisions related to village management. This not only facilitates the acceptance of the project but also fosters a strong sense of belonging and responsibility.

3.2.2.4 Project Monitoring Phase

Definition of success indicators: Monitoring should begin with the definition of clear indicators that measure the impact and effectiveness of the actions implemented. These indicators must be specific, measurable, and aligned with the initial project objectives. Some examples of indicators include improving access to digital services, the level of community participation, and the efficiency in managing natural resources.

Monitoring tools and data collection: To monitor the evolution of the project, advanced data collection tools are necessary. The use of digital platforms, *IoT* (Internet of Things) sensors for Cloud, and mobile applications enables the collection of real-time data on various aspects of the village, such as energy consumption, air quality, waste management, and civic participation. These data must then be analysed to adjust and optimize ongoing policies.

Feedback and participatory evaluation: A fundamental aspect of monitoring is the continuous involvement of the community. Citizens should be engaged in the evaluation process through surveys, online forums, and public meetings. This *feedback* process allows for quickly addressing any issues and adapting solutions to emerging needs.

3.2.2.5 Adaptation and Corrections

Monitoring is not a static activity, but a continuous cycle that requires adaptability. Based on the data collected and community evaluations, it is possible to review and modify the strategies and actions taken to ensure that the objectives are achieved effectively and sustainably. The implementation and monitoring of a *Smart Village* project are two closely interconnected phases that allow transforming the vision into reality. The initial planning must be followed by concrete actions, with a strong focus on community participation, the use of technology, and sustainability. Continuous monitoring and the adaptation of strategies are essential to ensure that the project remains relevant and responds to the needs of the community and the territory. Therefore, the *business plan* is not just a strategic document, but an operational guide that enables the realization of a sustainable, inclusive Smart Village, in tune with its territorial vocation. In addition to an excellent Business Plan, it is also necessary to develop a management protocol that considers how the various components should interface with each other. This will be elaborated in the next section.

3.2.3 Management Protocol for the Development of Smart Villages

To achieve a Smart Village, it is essential to follow a path inspired by the Smart Cities model but adapted to the specific characteristics and needs of rural contexts. This process is structured into phases that consider parameters, objectives, and collaborative networks (Fig. 3.5). The creation of a *Smart Village* requires an integrated strategy that includes multiple levels of action. The collaborative network [108] is like that envisioned for Smart Cities: local and regional governments, businesses, universities, and the community. The government plays a crucial role in providing the necessary *governance* and promoting policies that support the adoption of technologies, social inclusion, and sustainability. Businesses, particularly those in the technology and infrastructure sectors, are essential for providing innovative solutions. Universities and research centres contribute through knowledge production, technological innovation, and training. The community, finally, must be actively involved in the transformation process, through participation in collective decisions, management of shared projects, and strengthening social capital.

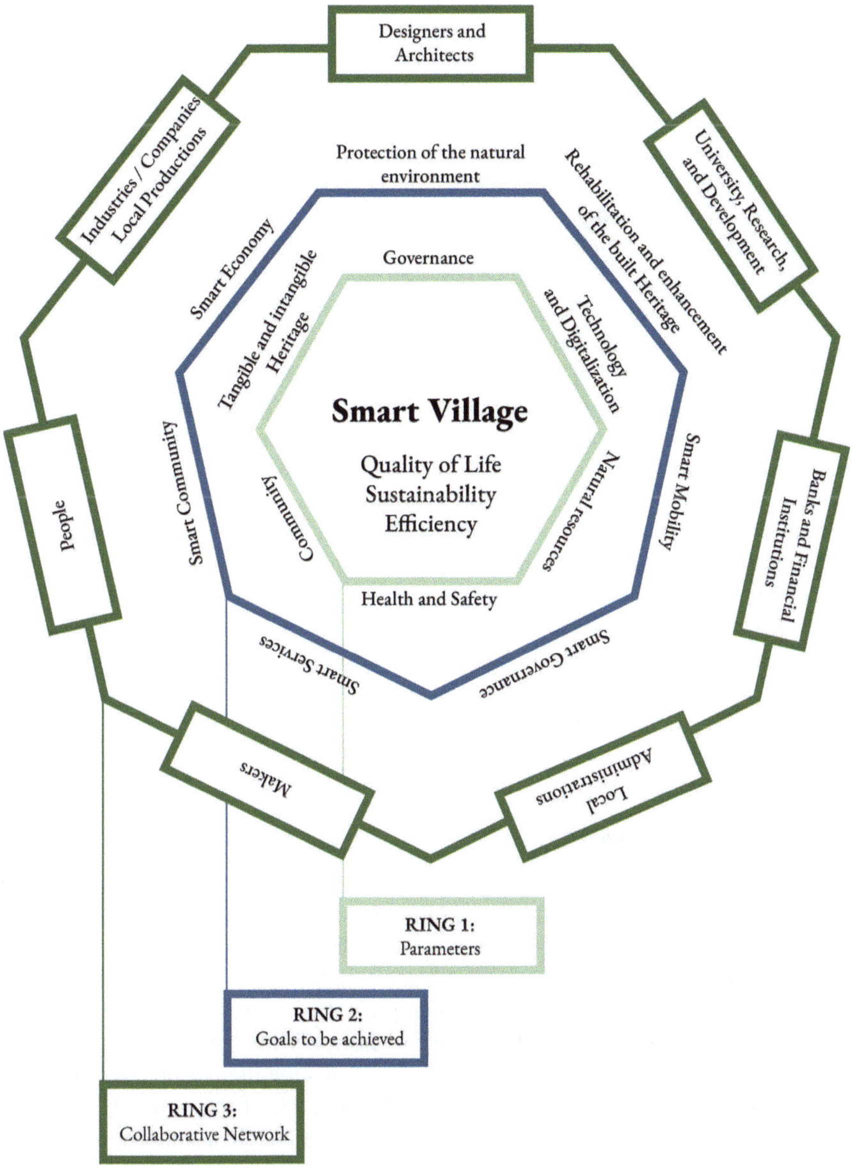

Fig. 3.5 Management protocol for smart villages. © Image by author, 2025

First, it is essential to develop a solid infrastructure that enables the management and interconnection of the various systems present in the *Smart Village* [109]. This involves, in addition to the aspects already considered in the collaborative network, health and safety indices and natural resources [110]. Collaboration between different actors, transparency in decision-making, and strengthening administrative capacities are fundamental aspects for the success of the project. These must then contribute to objectives to be achieved, namely ring 2. To make rings 1 and 3 work, a collaborative network is needed to ensure that the actors of level 1 and the objectives of level 3 are reached. There are no rigid time limits for creating a Smart Village. However, the transition must occur gradually, responding to the emerging needs of the area. The ideal moment to start the process is when the community and institutions are ready to undertake a path of change [111], with a strong will to innovate and improve living conditions. The context must also be favourable, with political support, economic resources, and growing awareness of the need for transformation.

Although the concept of a *Smart Village* is designed for rural areas, each context has its peculiarities that influence the transformation process. The size of the community, existing infrastructure, level of technological development, and support from local institutions are factors that determine the feasibility and speed of the project. Areas with a good level of digitalisation and strong connections with European or national rural development policies are more likely to initiate and successfully sustain a *Smart Village* project [112]. The path towards creating a Smart Village, therefore, requires a holistic approach that integrates technology, sustainability, and community. As with smart cities, *Smart Villages* also require integration between smart infrastructure, efficient *governance*, and strong community involvement. Only through cooperation between the public, private sectors, and citizens will it be possible to achieve lasting and meaningful transformation, contributing to the well-being of people and environmental sustainability. The result will be a more resilient, innovative rural community, prepared to face the challenges of the future.

3.3 A Smart Network

Rural areas, composed of multiple communities and municipalities spread across diverse landscapes, represent microcosms where history, nature, culture, and tradition intertwine with the challenges of modernity. The creation of a Smart Network for rural territories begins as an ambitious project capable of promoting sustainable and integrated development based on collaboration and innovation. At the heart of it lies the principle of mutual aid: stronger municipalities or communities can act as a driving force for weaker ones, yet the strength of the network lies in its ability to work together toward a common goal: sustainability and the realization of the SMART concept, particularly Smart Villages. The construction of a Smart Network [113] cannot be separated from the active involvement of the local community, which must feel engaged and proactive in driving change. Inhabitants are bearers of knowledge, experiences, and traditions that represent a fundamental resource for the development and future of rural areas. Participation must occur through moments of exchange, co-design workshops, and association initiatives—forms of participatory democracy including assemblies, meetings, workshops, and focus groups dedicated to specific themes—where ideas emerge from the bottom up and transform into concrete projects. Through this participation, rural areas can express their identity, enhance their specificities, and generate a sense of belonging. The governance of a Smart Network must be inclusive, transparent, and effective in seizing funding and support opportunities [103]. Through shared strategic planning, local administrations can access regional, national, and European funds, promoting interventions that respond to the specific needs of rural areas. A collaborative approach involving public and private entities and citizens allows for maximizing available resources and ensuring sustainable and long-term interventions. The cultural and natural heritage of rural territories represents a key element for revitalization. Protecting and rehabilitating architectural and landscape heritage are not only conservation efforts but also opportunities to stimulate the economic and social fabric, promoting both tradition and innovation. Similarly, intangible traditions—such as folk festivals, artisanal knowledge, religious celebrations, and local narratives—can be enhanced through events, tourist routes, and collective storytelling projects, turning the past into a resource for the future.

New technologies offer powerful tools to address the challenges of rural areas. Monitoring systems, sustainable energy management of built heritage, digital platforms for tourism and civic participation, applications for heritage enhancement, and support for local commerce are just a few of the possible solutions. Digitalisation can also improve access to essential services, reducing the gap between rural and urban centres [103]. The future of rural areas lies in the integration of innovative technologies that respect local identity. Smart Networks need to balance tradition and innovation. Projects must be rooted in local history and culture while remaining open to new ideas and technologies. For example, rehabilitating a small historic town can combine traditional materials and techniques with technological solutions for energy efficiency and connectivity. This approach enhances local resources while stimulating new development opportunities. To support the smart network and ensure that rural communities have access to the necessary resources, it is essential to have appropriate tools. The Rural Toolkit proposed by Europe serves as a comprehensive guide to European Union funding and support opportunities for rural areas, including direct applications, participation through national or regional authorities, and financial instruments. This guide provides detailed information on calls for proposals, programs, and European initiatives, offering concrete support for designing and implementing interventions. A prepared and well-informed territory is a territory that can face future challenges with greater awareness and determination [114]. However, it is useful to start a thoughtful and conscious path that systematizes projects and processes in a sustainable and virtuous way. While the EU-promoted toolkit serves as an aid for Smart Villages, this research aims to initiate a programmatic approach focused on identifying local specificities, rather than providing a mere collection of links for rural areas. The goal is to provide adequate tools that consider the true essence of territories, going beyond generic contexts and recognizing the uniqueness of each. A Smart Network for rural areas is not just a project but a development vision aimed at transforming the territory into an example of sustainability, innovation, and cooperation. Through mutual aid and inclusive governance, rural areas can become virtuous models, demonstrating that the strength of communities and the integration of past and future are keys to lasting success.

References

1. Scoones, I. (2015). *Sustainable livelihoods and rural development.* Fernwood Publishing.
2. Fraser, N. (2008). *Scales of justice.* Polity Press.
3. Fraser, N. (2018). Recognition without ethics? In C. McKinnon, & D. Castiglione (Eds.), The culture of toleration in diverse societies (pp. 21–42). Manchester University Press. https://doi.org/10.7765/9781526137708.00011
4. Fraser, N. (2019). *The old is dying and the new cannot be born.* Verso.
5. Chambers, R. (1983). *Rural development: Putting the last first.* Prentice Hall.
6. Duflo, E. (2012). *Human values and the design of the fight against poverty.* Tanner Lectures on Human Values.
7. Meloni, B., & Uleri, F. (2021). Ruralità e sviluppo: Il ruolo del turismo rurale oggi. *Dialoghi Mediterranei, 50*, 1–15.
8. Barca, F., Carrosio, G., & Lucatelli, S. (2018). Le aree interne da luogo di disuguaglianza a opportunità per il paese: Teoria, dati, politica. *Rivista di Politica Economica, 108*(3), 45–72.
9. DeFilippis, J. (2009). Paradoxes of community-building: Community control in the global economy. *International Social Science Journal, 59*(192), 223–234. https://doi.org/10.1111/j.1468-2451.2009.00694.x
10. Manzini, A. (2023). ELP. Sellerio, Palermo.
11. Mcmillan, D., & Chavis, D. (1986). Sense of community: A definition and theory. *Journal of Community Psychology, 14*, 6–23. https://doi.org/10.1002/1520-6629(198601)14:13.0.CO;2-I
12. Olivetti, A. (2021). *L'ordine politico delle comunità.* Edizioni di Comunità.
13. Cocco, A. (2022). The role of energy communities in the energy transition. *The Italian Law Journal.*
14. Mori, P. A., & Sforzi, J. (Eds.). (2018). *Imprese di comunità: Innovazione istituzionale, partecipazione e sviluppo locale.* Il Mulino.
15. Delle Cave, G. (2024). Comunità intelligenti, mobilità sostenibile e potere pubblico. Percorsi di diritto amministrativo—Monografie, 84. Franco Angeli.
16. Lombardo, L., Colajanni, S., & Campisi, T. (2024). NZED—Nearly Nero energy district at Polizzi Generosa: A village of the inner mountain Madonie Park in Sicily. *Sustainable Mediterranean Construction, 19*, 79–86.
17. Ahmadi, M., Kazemi-Arpanahi, H., Nopour, R., & Shanbehzadeh, M. (2023). Factors influencing quality of life among the elderly: An approach using logistic regression. *Journal of Education Health Promotion, 12*, 215. https://doi.org/10.4103/jehp.jehp_13_23
18. Darwin, S., Fischer, W., & Chesbrough, H. (2020). How to create *smart villages*: Open innovation solutions for emerging markets. Peaceful Evolution Publishing.
19. Iaione, C., De Nictolis, E., & Santagati, M. E. (2022). Participatory governance of culture and cultural heritage: Policy, legal, economic insights from Italy. *Frontiers in Sustainable Cities, 4*, 777708. https://doi.org/10.3389/frsc.2022.777708
20. Armeni, C., & Lee, M. (2021). Participation in a time of climate crisis. *Journal of Law and Society, 48*, 549–572. https://doi.org/10.1111/jols.12320
21. Donato, F., & Lohrasbi, A. (2017). When theory and practice clash: Participatory governance and management in Takht-e Soleyman. *Journal of Cultural Heritage Management and Sustainable Development, 17*, 129–146. https://doi.org/10.1108/JCHMSD-06-2015-0021
22. Alfonso Gonzales, P. (2018). Conceptualizing cultural heritage as a commons. In P. F. Biehl (Ed.),*Identity and heritage* (p. 30).

23. Donders, Y. (2018). Protection and promotion of cultural heritage and human rights through international treaties: Two worlds of difference? In C. Waelde, C. Cummings, M. Pavis, & H. Enright (Eds.), *Research handbook on contemporary intangible cultural heritage*. Edward Elgar Publishing.
24. Mottese, E. (2019). La Convenzione di Faro e la governance partecipativa del patrimonio culturale, Fogli di lavoro per il diritto in-ternazionale.
25. Watson, S., & Waterton, E. (2010). Heritage and community engagement. *Journal of Heritage Studies, 16*, 1–2. https://doi.org/10.1080/13527250903441655
26. Carayannis, E. G., & Campbell, D. F. J. (2010). Triple helix, quadruple helix and quintuple helix and how do knowledge, innovation and the environment relate to each other? A proposed framework for a transdisciplinary analysis of sustainable development and social ecology. *International Journal of Social Ecology and Sustainable Development, 1*, 41–69. https://doi.org/10.4018/jsesd.2010010105
27. Giuseppe, M. (2021). Path for new institutions: Urban commons in Naples. In *CIVIC eSTATE: New models of urban co-governance based on the commons*. Civic eState Final Network Product. Retrieved April 7, 2022, from https://urbact.eu/sites/default/files/final_product_-_civic_estate_new_models_of_urban_cogovernance_based_on_the_commons.pdf
28. Gould, P. G. (2018). *Empowering communities through archaeology and heritage: The role of local governance in economic development*. Bloomsbury Publishing.
29. Micciarelli, G. (2022). Hacking the legal. The commons between the governance paradigm and inspirations from the 'living his-tory' of collective land use. In F. Antònio, S. Federi-co, & C. Kim (Eds.), *Post-growth planning: Toward an urbanization beyond the market economy*. Routledge.
30. Steen, M. (2013). Co-design as a process of joint inquiry and imagination. *Design Issues, 29*, 16–28. https://doi.org/10.1162/DESI_a_00207
31. Yu, Y., Appiah, D., Zulu, B., & Adupoku, K. (2024). Integrating rural development, education, and management: Challenges and strategies. *Sustainability, 16*, 6474. https://doi.org/10.3390/su16156474
32. United Nations. (2023). *The sustainable development goals report 2023*. United Nations.
33. Ashley, C., Maxwell, S. (2001). Rethinking rural development. *Development Policy Review, 19*, 395–425.
34. Rachele, R. (2022). A long-term vision for the EU's rural areas. *Journal of European Rural Development, 28*, 15–34.
35. Scheyvens, R. (2002). *Tourism for development: Empowering communities*. Prentice Hall.
36. Nyaupane, G. P., Morais, D. B., & Dowler, L. (2006). The role of community involvement and number/type of visitors on tourism impacts: A controlled comparison of Annapurna, Nepal and Northwest Yunnan, China. *Tourism Management, 27*(6), 1373–1385. https://doi.org/10.1016/j.tourman.2005.12.013
37. Hernández, M., Vaquero, M., & Yubero, C. (2017). Cultural heritage and urban tourism: Historic city centres under pressure. *Sustainability, 9*, 1346. https://doi.org/10.3390/su9081346
38. Goodwin, H., & Santilli, R. (2009). *Community-based tourism: A success? ICRT Occasional Paper No. 11* (p. 37). International Centre for Responsible Tourism. https://www.haroldgoodwin.info/uploads/CBTaSuccessPubpdf.pd
39. Tricarico, L., & Geissler, J.-B. (2017). The food territory: Cultural identity as local facilitator in the gastronomy sector, the case of Lyon. *City, Territory and Architecture, 4*(1), 1–9. https://doi.org/10.1186/s40410-017-0076-8
40. Capello, R., & Perucca, G. (2017). Understanding citizen perception of European Union cohesion policy: The role of the local context. *Regional Studies, 52*(11), 1451–1463. https://doi.org/10.1080/00343404.2017.1397273
41. Buhalis, D., & Amaranggana, A. (2015). Smart tourism destinations enhancing tourism experience through personalization of ser-vices. In *Information and communication technologies in tourism 2015, ENTER 2015, Proceedings of the international conference in Lugano, Switzerland, February 3–6, 2015* (pp. 377–89). https://doi.org/10.1007/978-3-319-14343-9_28

42. Borzaga, C., Calzaroni, M., Carini, C., & Lori, M. (2019). *Italian cooperatives: An analysis of their economic performances, employment characteristics and innovation processes based on combined use of official data* (CIRIEC Working Paper No. 1906). CIRIEC—Université de Liège.
43. Lombardo, L. (2023). Energy communities and smart villages in the Madonie Sicilian inner rural area. In A. Sayigh (Ed.), *Mediterranean architecture and the green-digital transition, Innovative Renewable Energy*. Springer.
44. Kumari, P. L. S., Sailaja, P., Keerthana, K. K., & Mohana, R. M. (2022). Smart villages-scope for IoT and cloud applications. In M. Bhu-shan, S. Iyer, A. Kumar, T. Choudhury, & A. Negi (Eds.), *Artificial intelligence for smart cities and villages: Advanced technologies, development, and challenges* (pp. 216–246). Bentham Science Publishers. https://doi.org/10.2174/9789815049251122010015
45. Lombardo, L. (2023). Mediterranean identity sites. Technological challenges and smart solutions in rural areas. *IN FOLIO, 41*, 115–125.
46. Sagyndykova, Z., Gazdiyeva, B., Gabdullina, Z., Akhmetzhanova, A., Mektepbayeva, D. A., & Althonayan, A. (2021). The role of summer intensive programmes on improving students' learning outcomes in rural areas. *International Journal of Emerging Technologies in Learning (iJET), 16*, 122. https://doi.org/10.3991/ijet.v16i06.18713
47. Lepore, F., Ferrari, A., Fiorentini, N., Mannari, C., Bacco, M., & Brunori, G. (2023). *Co-design and e-governance tools for sustainable land and water management in rural areas: The experience within the DESIRA H2020 project* (pp. 25–30). https://doi.org/10.1109/MetroAgriFor58484.2023.10424263
48. Jones, E. (2025, February). *Co-design a school community*. Ellis Jones. https://www.ellisjones.com.au/blog/co-design-a-school-community
49. Phillips, R., Harper, S., & Gamble, S. (2007). Summer programming in rural communities: Unique challenges. *New Directions for Youth Development, 2007*, 65–73. https://doi.org/10.1002/yd.213
50. Jitea, I. M., Mihai, C. V., & Dumitras, D. E. (2013). The role of local actors' in identifying the needs for sustainable HNV farming. In *Dealurile Clujului Est. Book of abstracts, 17th international symposium "Prospects for the 3rd Millennium Agriculture", Cluj-Napoca, September 2018.*
51. Pantovic, D., Cvijanovic, D., Cvijanović, G., & Šobić, L. (2023). Tradition and culture as the base for the tourism product: Case of UNESCO intangible heritage. *Facta Universitatis, Series: Economics and Organization, 103.* https://doi.org/10.22190/FUEO230130007P
52. Inzerillo, B., & Morvillo, S. (2024). Food, design, and territory: The valorization of manna in the Madonie area. In *Food, design, and territory: The valorization of manna in the Madonie area* (pp. 635–642). Palermo University Press.
53. Lee, S. (2024). Resources and territorial claims: Domestic opposition to resource-rich territory. *International Organization, 78*(3), 361–396. https://doi.org/10.1017/S0020818324000134
54. Knapp, J., Zeratsky, J., & Kowitz, B. (2022). *Sprint. Come risolvere grandi problemi e testare nuove idee in soli cinque giorni.* Edizioni LSWR.
55. Abdulai, A. et al. (2020). Smart villages: A systematic literature review. *Sustainability, 12*(24), 10482.
56. Alsharif, M. H., & Drew, S. (2010). *E-Government adoption in the developing world: A theoretical.*
57. Stojanova, S., Lentini, G., Niederer, P., Egger, T., Cvar, N., Kos, A., & Stojmenova, E. (2021). Smart villages policies: Past, present and future. *Sustainability, 13*, 1663. https://doi.org/10.3390/su13041663
58. Pazos-Vidal, S., Vernis, A., & Tomey, M. (2021). Rural proofing ruralizar las leyes. Una cuestión de justicia. https://doi.org/10.13140/RG.2.2.27973.58087
59. Gupta, H., & Jana, R. K. (2017). E-Governance in rural development: A review of literature. *Journal of Rural Development, 36*(4), 501–517.
60. Donolo, R., & Ariccio, S. (2020). La resilienza sociale post-sisma delle comunità nelle aree interne delle Marche: Proposta di un metodo sistematico per misurarla, valorizzarla e favorirne

l'inclusione nella pianificazione e nelle politiche territoriali (parte I). Scienze Regionali. 10. DPCM 21 Dicembre 2016. Approvazione del Piano nazionale per lo sviluppo e la competitività delle aree interne.
61. Ebrahimbaysalami, O., & Ren, X. (2024). Heritage in transition: Vernacular architectural patterns in rural Iran. *Heritage, 7*(7), 3393–3416. https://doi.org/10.3390/heritage7070160
62. Lombardo, L., & Morvillo, S. (2024). Reviving from Ruins Sicilian sustainable development design through the reclamation of abandoned architectures. *IN FOLIO, 43*, 199–207.
63. Münch, A., Gorny, H., Badouix, M., Gaugitsch, R., Dwyer, J. C., Kubinakova, K., Beck, M., Bunnen, P. V., Mantino, F., & Brkanovic, S. (2024). European Commission: Directorate-general for agriculture and rural development. In *Study on funding for EU rural areas—Final report*. Publications Office of the European Union. https://doi.org/10.2762/901111
64. Mack, G., Fintineru, G., & Kohler, A. (2020). Effects of EU rural development funds on newly established enterprises in Romania's rural areas. *European Planning Studies, 29*(2), 291–311. https://doi.org/10.1080/09654313.2020.1746243
65. Claval, P. (2012). Changing conception of heritage and landscape. In N. Moore, & Y. Whelan (Eds.), *Heritage, memory and the politics of identity: New perspectives on the cultural landscape* (pp. 85–94). Ashgate Publishing, Hampshire.
66. ICOMOS. (1999). *Charter on the built vernacular heritage*. World Cultural Center.
67. Sardaro, R., la Sala, P., de Pascale, G., & Faccilongo, N. (2021). The conservation of cultural heritage in rural areas: Stakeholder preferences regarding historical rural buildings in Apulia, Southern Italy. *Land Use Policy, 109*, Article 105662. https://doi.org/10.1016/j.landusepol.2021.105662
68. Yang, Y., Zhang, H., Chen, M., Jiang, Y., & Chai, H. (2021). An inheritance mode of rural cultural heritage based on virtual museum in China. *International Journal of Computer Games Technology, 2021*, 4787991, 14 pages. https://doi.org/10.1155/2021/4787991
69. Yang, Y., Bai, Z., Zhang, H., & Wang, Y. (2023). The construction and application of a cloud editing digital museum oriented to virtual tour. *International Journal of Computer Games Technology, 2023*, 7132476, 13 pages. https://doi.org/10.1155/2023/7132476
70. Sousa, B., Santos, R., & Lubowiecki-Vikuk, A. (2021). Slow tourism as a tourism alternative to sustainable development. *Journal of Environmental Management and Tourism, 5*, 1404–1409. https://doi.org/10.14505/jemt.v12.5(53).24
71. Salemink, K., Strijker, D., & Bosworth, G. (2017). Rural development in the digital age: A systematic literature review on unequal ICT availability, adoption, and use in rural areas. *Journal of Rural Studies, 54*, 360–371. https://doi.org/10.1016/j.jrurstud.2015.09.001
72. Moscarelli, R., Toma, G., & Introini, V. (2023). Aree interne e turismo lento: un'analisi sulla coerenza del-le politiche nazionali lungo la Via Francigena e la Ciclovia dell'Acquedotto Pugliese. *Territorio*, 129–139. https://doi.org/10.3280/TR2022-101016
73. Galderisi, A., Fiore, P., & Pontrandolfi, P. (2021). Strategie operative per la valorizzazione e la resilienza delle aree interne: il progetto ri. *PRO VA. RE. BDC. Bollettino Del Centro Calza Bini.*
74. Onesti, A. (2013). Il recupero edilizio nell'approccio del paesaggio storico urbano. Gli strumenti per condividere le regole. *BDC. Bollettino Del Centro Calza Bini.* https://doi.org/10.6092/2284-4732/2457
75. Borri, A., Avorio, A., & Cangi, G. (2001). Riparazione e consolidamento degli edifici in muratura. In *Aa.Vv., Manuale per la riabilitazione e la ricostruzione postsismica degli edifici* (pp. 227–287). DEI Tipografia del Genio Civile.
76. Pinto, M. R. (2004). *Il riuso edilizio. Procedure, metodi ed esperienze*. UTET.
77. Marzano, G. (2012). L'alternativa comunitaria degli ecovillaggi: La relazionalità pragmatica di Torri Superiore. *Rivista di Scienze Sociali.*
78. Chioni, C., & Di Leo, B. (2024). *Rappresentazione digitale per la progettazione delle aree interne italiane.*
79. Buhmann, E., Ervin, S., Fricker, P., Hehl-Lange, S., & Palmer, J. (Eds.). (2016–2023). *Journal of Digital Landscape Architecture*. Retrieved January 10, 2024, from https://gis-point.de/jodla.html

80. Ferretti, M., Favargiotti, M., Lino, B., & Rolando, D. (2022). Branding4Resilience: Explorative and collaborative approaches for inner territories. *Sustainability, 14*(18), 11235. https://doi.org/10.3390/su141811235
81. Chioni, C., Maragno, A., Pianegonda, A., Ciolli, M., Favargiotti, S., & Massari, G. A. (2023). Low-cost 3D virtual and dynamic reconstruction approach for urban forests: The Mesiano University Park. *Sustainability, 15*(19), 14072. https://doi.org/10.3390/su151914072
82. Arkhipova, O., Semenova, O., Afanasenko, V., & Podoyan, E. (2021). Monitoring of historical and cultural heritage objects based on modern information technologies. *InterCarto. InterGIS, 27*, 434–444. https://doi.org/10.35595/2414-9179-2021-4-27-434-444
83. Liew, J.-H., Yeo, A., Hamid, K., & Othman, A.-K. (2004). *Implementation of wireless networks in rural areas* (p. 5).
84. Cisco Systems, Inc. (2004). *Data sheet: Cisco Aironet 350 series access points*. Retrieved April 5, 2004, from http://www.cisco.com/en/US/products/hw/wireless/ps458/products_data_sheet09186a008009247c.html
85. Wood, J. (2000). A standard approach to control room design. *Proceedings of the Human Factors and Ergonomics Society Annual Meeting, 44*, 6–417. https://doi.org/10.1177/154193120004403505
86. Barrientos, F., Martin, J., DE Luca, C., Tondelli, S., Gómez-García-Bermejo, J., & Zalama Casanova, E. (2021). Computational methods and rural cultural and natural heritage: A review. *Journal of Cultural Heritage, 49*, 250–259. https://doi.org/10.1016/j.culher.2021.03.009
87. Bertolotti, E. (2019). *Local communities have stories to tell: Storytelling for social inclusion.* https://doi.org/10.1007/978-3-030-31117-9_7
88. Agnoletti, M. (2014). Rural landscape, nature conservation and culture: Some notes on research trends and management approaches from a (southern) European perspective. *Landscape and Urban Planning, 126*, 66–73. https://doi.org/10.1016/j.landurbplan.2014.02.012
89. Nisticò, R. (2019). L'Italia da riabitare, oltre la trappola della marginalità. *Rivista Economica del Mezzogiorno, 33*(3–4). https://doi.org/10.1432/96258
90. Di Bitonto, P., Di Tria, F., Laterza, M., Roselli, T., Rossano, V., & Tangorra, F. (2010). Automated generation of itineraries in recommender systems for tourism. In F. Daniel, & F. Facca (Eds.), *ICWE 2010 workshops, Lecture notes in computer science* (Vol. 6385, pp. 498–508). Springer-Verlag Berlin. https://doi.org/10.1007/978-3-642-16985-4_48
91. Hopkins, J., & Copus, A. (2017). *Web-based mapping for open access—Building capacity and exploring user preferences*. Retrieved July 19, 2018, from http://www.hutton.ac.uk/research/projects/web-based-mapping-openaccess-building-capacityand-exploring-user-preferences
92. Trapp, M., Semmo, A., Pokorski, R., Herrmann, C., Döllner, J., Eichhorn, M., & Heinzelmann, M. (2010). Communication of digital cultural heritage in public spaces by the example of Roman Cologne. In M. Ioannides, D. Fellner, A. Georgopoulos, & D. Hadjimitsis (Eds.), *EuroMed 2010: Digital heritage, Lecture notes in computer science* (Vol. 6436, pp. 262–276). Springer Berlin Heidelberg. https://doi.org/10.1007/978-3-642-16873-4_20
93. Mao, Z., Zhu, X., Zou, Q., & Jin, W. (2024). How can digital villages improve basic public services delivery in rural areas? Evidence from 1840 counties in China. *Agriculture, 14*, 1–20. https://doi.org/10.3390/agriculture14101802
94. Mukhsin, M., Rizky, A., & Anggreni, L. (2018). The study of e-government implementation in rural areas. In *Proceedings of ICOSAPS-18*. https://doi.org/10.2991/icosaps-18.2018.31
95. Stein, V., Pentzold, C., Peter, S., & Sterly, S. (2022). Digitalisation and civic participation in rural areas. A systematic review of scientific journals, 2010–2020. *Raumforschung und Raumordnungspatial Research and Planning, 80*(3), 251–265. https://doi.org/10.14512/rur.112
96. Moseley, M. J. (1979). *Accessibility: The rural challenge*. Methuen.
97. Farrington, J., & Farrington, C. (2005). Rural accessibility, social inclusion and social justice: Towards conceptualisation. *Journal of Transport Geography, 13*, 1–12.
98. da Silva, F. M., & Almendra, R. (2007). Inclusive design: A new approach to design project. In M. S. Pereira (Ed.), *A portrait of state-of-the-art research at the Technical University of Lisbon.* Springer. https://doi.org/10.1007/978-1-4020-5690-1_37

99. Mikou, M., Rozenberg, J., Koks, E., Fox, C., & Peralta Quiros, T. (2019). *Assessing rural accessibility and rural roads investment needs using open source data* (Policy Research Working Paper No. 8746). © World Bank. http://hdl.handle.net/10986/31309
100. Mańkowska, D., Szałata, Ł., & Derlukiewicz, N. (2024). *Managing rural development towards smart village*. Przegląd Organizacji. https://doi.org/10.33141/po.2023.03.17
101. Letaifa, S. B. (2015). How to strategize smart cities: Revealing the SMART model. *Journal of Business Research, 68*(7), 1414–1419. https://doi.org/10.1016/j.jbusres.2015.01.024
102. Stoustrup, S. W. (2021). The re-coding of rural development rationality: Tracing EU Governmentality and Europeanisation at the local level. *European Planning Studies, 30*(12), 2474–2491. https://doi.org/10.1080/09654313.2021.2009776
103. Zavratnik, V., Kos, A., & Stojmenova, E. (2018). Smart villages: Comprehensive review of initiatives and practices. *Sustainability, 10*, 2559. https://doi.org/10.3390/su10072559
104. European Parliament legislative resolution on the proposal for a Council decision on Community strategic guidelines for rural development (Programming period 2007–2013) (COM(2005)0304—C6-0349/2005—2005/0129(CNS)).
105. European Union. (2023). *Preparatory action—Smart rural areas in the 21st century: Executive summary*. Publications Office of the European Union. https://doi.org/10.2762/72489
106. Paiola, M., Grandinetti, R., Kowalkowski, C., & Rapaccini, M. (2024). Digital servitization strategies and business model in-novation: The role of knowledge-intensive business services. *Journal of Engineering and Technology Management, 74*, 101846. https://doi.org/10.1016/j.jengtecman.2024.101846
107. Rosati, U., & Conti, S. (2016). What is a smart city project? An urban model or a corporate business plan? *Procedia - Social and Behavioral Sciences, 223*, 968–973. https://doi.org/10.1016/j.sbspro.2016.05.332
108. Albino, V., Berardi, U., & Dangelico, R. M. (2015). *Smart cities: Definitions, dimensions.*
109. Barile, S., Pellicano, M., & Polese, F. (Eds.). (2018). *Social dynamics in a systems perspective.* Springer International Publishing.
110. Batagan, L. (2011). Smart cities and sustainability models. *Revista de Informatica Economica, 15*(3), 80–87.
111. Ciasullo, M. V., Troisi, O., Grimaldi, M., & Leone, D. (2020). Multi-level governance for sustainable innovation in smart communities: An ecosystems approach. *International Entrepreneurship and Management Journal*, 1–29.
112. Komninos, N. (2006). The architecture of intelligent cities: Integrating human, collective and artificial intelligence to enhance knowledge and innovation. In IEEE 2nd IET international conference on intelligent environments (pp. 13–20). IEEE Xplore.
113. Emerllahu, V., & Bogataj, D. (2024). Smart villages as infrastructure of rural areas: Literature review and research agenda. *IFAC-PapersOnLine, 58*(3), 268–273. https://doi.org/10.1016/j.ifacol.2024.07.162
114. Satola, L., & Milewska, A. (2022). The concept of a smart village as an innovative way of implementing public tasks in the era of instability on the energy market examples from Poland. *Energies, 15*(14), 5175.

